"This book, written by two of the most important scholars in the environmental social sciences today, could not be more timely. Breaking down the social mechanisms that shape processes of change in our socio-ecological relations, and written in an accessible manner, this book has something for students, seasoned researchers, and concerned publics alike. Rather than hand waving about the monstrous environmental and climate crises we face, Environmental Sociology and Social Transformation invites readers to roll up their sleeves and get to work."

Debra Davidson, *University of Alberta, Canada*

"This imaginative and important book brings sociological critique and transformative change together. Boström and Lidskog explain how environmental problems are societal problems – with social causes and social solutions. A useful text for students, activists and policy makers seeking to understand and solve our most pressing global issues."

Jo Lindsay, *Monash University, Australia*

"This book is a remarkable addition to environmental sociology debates, stressing the need to understand society's role in contemporary ecological challenges. By superbly presenting the 'five facets of the social' it unravels the complexities of our civilizational crisis and provides a compass for just social transformation. This contribution is pertinent to audiences worldwide."

Llewellyn Leonard, *University of South Africa*

"In this volume, Boström and Lidskog offer a wide-ranging and accessible overview of the diverse perspectives and key concepts that show why a sociological lens is so vital for understanding environmental issues, impacts and solutions. By consistently making the connections between everyday practices and more global phenomena and issues, this is an ideal book to help students navigate the complex ways in which 'global' environmental problems are intertwined with our everyday lives."

Mark C.J. Stoddart, *Memorial University, Canada*

"As the environmental crisis becomes increasingly worrisome and encompassing, this book offers an invaluable compass for navigating its many facets, making sense of its social causes and possible solutions."

Luigi Pellizzoni, *University of Pisa, Italy*

"Without understanding society, we cannot properly address the environmental crisis we are facing today. That is where sociological theory comes in, and also this well-written book. It will be a very enlightening work for scholars and students seeking to learn about the environmental crisis and what we can do about it."

Saburo Horikawa, *Hosei University, Japan, and President, the Japanese Association for Environmental Sociology*

"In response to escalating planetary crisis, Boström and Lidskog urgently explore the imperative for profound social transformation, employing sociological theory to dissect environmental challenges. This eloquently written book is a tour de force that will shape the academic and policy discourse for years to come."

Audronė Telešienė, *Kaunas University of Technology, Lithuania*

Environmental Sociology and Social Transformation

Environmental Sociology and Social Transformation demonstrates how sociological theory and research are critical for understanding the social drivers of global environmental destruction and the conditions for transformative change.

Written by two professors of sociology who are deeply involved in the international community of environmental sociology, Magnus Boström and Rolf Lidskog argue that we need to better understand society as well as the fundamentally social nature of environmental problems and how they can be addressed. The authors provide answers to why so many unsustainable practices are maintained and supported by institutions and actors despite widespread knowledge of their negative consequences. Employing a pluralistic sociological approach to the study of social transformations, the book is divided into five key themes: Causes, Distributions, Understandings, Barriers, and Transformation. Overall, the book offers an integrative and comprehensive understanding of the social dimension of (un)sustainability, societal inertia, and conditions for transformative change. It provides the reader with references from classic and contemporary sociology and uses pedagogical features including boxes and questions for discussion to help embed learning.

Arguing that a broad and deep social transformation is needed to avoid a global civilization crisis, *Environmental Sociology and Social Transformation* will be a great resource for students and scholars who are exploring current environmental challenges and the societal conditions for meeting them.

Magnus Boström is Professor of Sociology at Örebro University, Sweden, with expertise in environmental sociology and the sociology of consumption. His current research focuses on the social drivers behind unsustainable mass consumption, as well as the conditions for lifestyle change and more collective ways of sharing resources.

Rolf Lidskog is Professor of Sociology at Örebro University, Sweden, with expertise in environmental sociology, environmental politics, and risk regulation. He is currently conducting research on international environmental governance, focusing on its conditions and the role of expertise.

Key Issues in Environment and Sustainability

This series provides comprehensive, original and accessible texts on the core topics in environment and sustainability. The texts take an interdisciplinary and international approach to the key issues in this field.

Sustainable Consumption
Key Issues
Lucie Middlemiss

Human Rights and the Environment
Key Issues
Sumudu Atapattu and Andrea Schaper

The Anthropocene
Key Issues for the Humanities
Eva Horn and Hannes Bergthaller

Environmental Justice
Key Issues
Edited by Brendan Coolsaet

Ecosystem Services
Key Issues
Mark Everard

Sustainable Business
Key Issues
Helen Kopnina, Rory Padfield and Josephine Mylan

For more information about this series, please visit: www.routledge.com/Key-Issues-in-Environment-and-Sustainability/book-series/KIES

Environmental Sociology and Social Transformation

Key Issues

Magnus Boström and Rolf Lidskog

LONDON AND NEW YORK

from Routledge

Designed cover image: Getty images

First published 2024
by Routledge
4 Park Square, Milton Park, Abingdon, Oxon OX14 4RN

and by Routledge
605 Third Avenue, New York, NY 10158

Routledge is an imprint of the Taylor & Francis Group, an informa business

British Library Cataloguing-in-Publication Data
A catalogue record for this book is available from the British Library

ISBN: 978-1-032-60655-2 (hbk)
ISBN: 978-1-032-60653-8 (pbk)
ISBN: 978-1-032-62818-9 (ebk)

DOI: 10.4324/9781032628189

Typeset in Times New Roman
by codeMantra

Contents

Figures

Boxes

Abbreviations

COP	Conference of the Parties
EU	European Union
IAEA	The International Atomic Energy Agency
IPBES	The Intergovernmental Science-Policy Platform on Biodiversity and Ecosystem Services
IMF	International Monetary Fund
IPCC	The Intergovernmental Panel on Climate Change
NETs	Negative Emission Technologies
SDG	Sustainable Development Goals
UN	United Nations
UNEP	United Nations Environment Programme
WHO	World Health Organization
WTO	World Trade Organization
WWF	World Wildlife Fund

1 Crisis

The need for social transformation

Our societies are in turmoil. On the one hand, extraordinary developments have occurred in the last few centuries of human history. These include dramatic increases in material wealth, improvements in public health, the development of democracy and human rights, and the lifting of people out of poverty. On the other hand, we face climate change, biodiversity loss, and pollution. These threats are no longer abstract or something that will happen in the future. The dark side of development has materialized and is constantly affecting people and places around the world. Societies and communities are struggling to cope with a growing number of problems. Some of these are creeping, such as deteriorating air quality, erosion, declining freshwater levels, and plastic pollution in the oceans. Others are acute, such as extreme weather and its consequences in the form of flooding, storms, and wildfires. When resources become scarcer, social conflicts follow, along with increasing migration, higher prices, and health problems. Scientists speak of a planetary crisis – planetary in the sense that the dangers no longer threaten only particular localities and regions, but societies worldwide. Risks are materializing into disasters.

This book acknowledges the environmental challenges facing humanity, which appear to be of unprecedented magnitude, at least for modern society. It does so with the central argument that in order to understand the crisis and what we can do about it, we need to understand society. This book shows how sociological theory and research are crucial for understanding the social causes of global environmental challenges and the conditions for transformative change.

This situation makes social transformation the most important and urgent issue of our time. But how is it possible to transform a society, or even a culture or civilization? And what exactly is it that needs to be changed? To develop realistic and viable guidance, it is important to consider not only where we are heading (if nothing is done), but also why we are heading there, that is, to examine the causes behind the current developments. Behind all global environmental challenges there are *social causes* that need to be understood, prevented, or redirected.

Sociological knowledge is essential to gaining a better understanding of these social causes and generating ideas for how transformative change can take place. Otherwise, science may continue to accumulate knowledge about the deteriorating environmental situation and testify to the growing need for society

DOI: 10.4324/9781032628189-1

to fundamentally change track – yet remain paralyzed when it comes to making these necessary changes. The actions taken so far have been insufficient, and the solutions proposed are inadequate. It is therefore necessary to address the more fundamental causes of environmental destruction. Moreover, to be able to transform society, we need to identify and investigate the barriers to change, and find ways to overcome them. If the barriers are not recognized and addressed, all our ambitions and talk about transforming society will only be wishful thinking: strong words on paper, but weak in practice. Thus, ambitions to change society must be based on a valid understanding of society and why it continues its unsustainable activities. Why are many unsustainable practices perpetuated and often even actively promoted by institutions and actors despite widespread knowledge of their negative consequences?

This book is based on two basic assumptions. The first is that society is facing a global environmental challenge of extreme magnitude. This challenge has increased not only in scope (global problems) and breadth (encompassing all sectors of society) but also in depth (concerning fundamental societal structures). This implies that moderate and gradual change is insufficient, and the solution is transformative change, that is, change that restructures society. It also makes a lot of sense to describe the current environmental challenge as a civilizational crisis. If nothing is done, it will seriously threaten humanity. Averting this crisis will require a transformation so profound that it will change society deeply and broadly.

The second assumption is that in order to be able to respond to this environmental challenge, it is necessary to understand its roots and to identify viable – relevant and realistic – ways forward. This places society at the center, because the global environmental crisis is caused by society and can only be solved by society. The need to change society makes it necessary to better understand how it has evolved historically and how it functions today. We should not locate these roots to any particular level or aspect of society, such as global forces or individual values. The scope, breadth, and depth of the environmental crisis indicate that its roots can be traced to all levels and aspects of society. We therefore argue that there is a great need to properly understand "the social," that is, society in all its complexity. If we seek social transformation, we must understand its conditions and why until now it has been so difficult to change society in a sustainable direction. Such knowledge also brings a dose of caution against overly simplistic demands such as "tear down everything now," or "replace our society immediately with a new one." History is full of failed attempts to change society, as well as successful attempts that turned out to have enormous unintended and negative consequences.

This book adopts a critical-constructive perspective. We develop a critical perspective, which means we believe that some established ways of thinking and acting, such as relying on increased economic growth and consumption as the way to prosperity, ought to be criticized. We also emphasize the need for a constructive perspective. It is not enough to criticize current society, policies, and development paths; it is also important to show or suggest viable ways forward. This constructive work requires reflexivty. Discussing, hoping, and working for a better future will also require a great deal of imagination, both to unpack current (and often implicit) notions of socio-ecological futures and to open up ways of seeing other possible futures and the ways to get there. A critical-constructive perspective also requires transparency about one's own normativity, that is, clarifying one's normative commitments and value assumptions. For example, if environmental measures and proposed solutions do not take into account social justice, with its stress on equity and quality of life, they may be extremely unfair and undemocratic.

The book structures its argument around five broad and overlapping thematic issues. A key argument of the book is that these five themes are inherently social; none of them can be reduced to an ecological, economic, technical, cultural, or political dimension. The five themes, to each of which a chapter is devoted, are:

1. Causes: the social roots of environmental problems
2. Distributions: the social spread of environmental problems
3. Understandings: the social sense-making of environmental problems
4. Barriers: social resistance through inadequate solutions
5. Transformations: ways of changing society

The order of the chapters does not imply a sequential line of reasoning. Causes, distributions, understandings, barriers, and transformations are not distinct, successive phases or stages of environmental problems. Instead, they are dynamically related and intertwined. They intersect and weave back and forth in unexpected and unpredictable ways.

We apply a pluralistic approach to the study of social transformations (Boström and Davidson 2018). We are also inspired by a sociological imagination approach (Mills 2000). Sociological imagination is the ability to shift perspectives, to see that circumstances could be different, and to view contemporary events as both part of history and something that can shape history. It also means seeing that what was unthinkable yesterday can be thinkable, possible, and even realized, today. Such a reflexive and imaginative capacity, which requires critical-constructive thinking, is arguably more important today than ever before. The current environmental challenge is of such unprecedented magnitude that it is not enough just to see that something is going wrong with our current way of life; we must also be able to perceive everything that is happening and work in the right direction, as well as to create the seeds of change. We will need a great deal of socio-ecological imagination to find, propose, and test ways forward.

No one can look at the world from a purely detached position, unaffected by any social and historical context. Our position influences how we are, how we see, and how we act in the world. This is true not only of people in general, but also of research. We must have a position (providing assumptions and perspective) in order to gain knowledge about the world. Of course, scientific training is of great help in revealing biases, questionable assumptions, and poorly formulated hypotheses and concepts. The fact that knowledge is always gained from a particular point of view, as philosophy of science has long shown, does not mean that there is no such thing as valid knowledge, but only that we must always reflect on the perspective from which knowledge has been gained.

Therefore, as the authors of this book, we cannot entirely disentangle ourselves from how our understanding of the world and its challenges is shaped by being white males living in a materially privileged part of the world. At the same time, this does not make us incapable of considering how our social circumstances and our socialization into the research community have affected our ways of understanding and gaining knowledge about the world. Human beings have a capacity for reflection, including self-reflection. Although we necessarily view society from a particular vantage point, we are able to observe, interpret, understand, critique, and reflect on how our experiences and positions influence our being and seeing.

As stated in this introduction, the fundamental assumption underlying this book is that we face an enormous global environmental threat to which society must respond. In short, *broad and deep social transformations* are needed *in order to avoid a global catastrophe (civilizational crisis) and instead achieve long-term sustainability*. In this introduction, we first present and develop our theoretical understanding of "the social," a theoretical model that we call "the five facets of the social." We then discuss the meaning of the concepts of civilizational crisis and transformation. We will then briefly introduce the five overarching themes that structure the coming chapters of the book: causes, distributions, understandings, barriers, and transformations.

The five facets of the social

"The social" is the most fundamental analytical category in sociology; indeed, it is even definitional for the discipline of sociology. An emphasis on the social is a common feature in the history of sociology, from its origins in the nineteenth century to the present. As early as the mid-nineteenth century, August Comte coined the term sociology, or social physics, stating that sociology aims to uncover the laws that govern social reality. Half a century later, Emile Durkheim noted that society can be explained not in terms of its individual members, but by social facts. Around the same time, Georg Simmel stressed that even if society is composed of individuals, it forms a "higher unity." Max Weber, another classical sociologist, stated that sociology is the science that understands the meaning of social action, that is, how an individual considers the possible behaviors of other actors as providing guidance for her own actions.

Nevertheless, and somewhat ironically, the question of what sociologists include in the category of the social is seldom explicitly discussed or debated. It is as if it were self-evident what sociologists and other social scientists mean, and where they focus their attention, when they speak of phenomena such as *social* movements, *social* relations, *social* structures, *social* order, *social* change, *social* integration, *social* capital, and so on.

The word "social" is also popular outside sociology. The sustainability discourse uses the term *social* sustainability, and in common language the word is often associated with sociality or the phrase "to be social." The broad and vague use of the word has led anthropologist Bruno Latour (2005) to argue that "the social" should no longer be used by social scientists. By using the prefix "social," they assume a certain character of an entity or process – in a similar way as when describing things as "biological," "economic," "mental," or "organizational." According to Latour, however, the social is not a dimension or domain of reality. Similarly, he considers using the idea of society in explanations to be like having a large container ship that no inspector is allowed to board, and no one knows what might be hidden in the cargo (ibid, p. 68). He finds *social* explanations to be too cheap, too automatic, and to contain so many ingredients that it has become impossible to unpack them (ibid, p. 221). We agree with Latour that "the social" always runs the risk of being used as a simple and superficial way of explaining complex processes, and when used in a vague sense it conceals rather than reveals what is happening and why. However, we also consider "the social" to be a crucial concept for understanding our lives and our world. As sociologists, we see it as a deep and fundamental phenomenon that is constitutive of human life. Although the aim of this book is not to provide a complete theory of the social, if that is even possible, we argue that it is necessary to understand *the social* in order to understand the conditions that determine how society can be transformed.

This means that the social cannot simply be seen as a "box" alongside other boxes (such as economics, politics, technology, and the environment) that together form a general model of society or reality. This way of understanding society is rather common in the sustainability literature and practice. As Boström and Davidson (2018, pp. 11–12) argue, there is often an element of ecological (or other) reductionism involved:

> In many environmental conceptualizations, the ecological domain is treated as something objectively fixed and given, something to which the social domain must be oriented. Perhaps more often the social is treated as a single variable, factor or pillar, or a limited list of elements, or a black box ... In other words, the social dimension is reduced to (just) one measurable and manageable aspect, such as "population" (to be controlled), "population growth" (to be stopped and potentially reversed), "attitudes" (to be changed), or "the public" (to be educated).

In contrast, human life is not conceivable or even possible without taking the social into account. The ways in which people live their lives – which, with few exceptions, are unsustainable – cannot be transformed unless we understand that the social pervades all aspects of human life and all aspects of society. Informed by sociological thinking, we understand the social as comprising five intertwined facets.

First, the social can be understood as part of our *inner life*, within our personality or identity, in social constructions (norms, conceptions, emotions) that we share with others (intersubjectively). These constructions are acquired through primary and secondary socialization (Berger and Luckmann 1967/1991) or through social interaction with significant others in one's social context (Mead 1934/1972, Blumer 1969). Even the solitary individual is social; thoughts, expectations, and feelings are developed in interaction with others and oriented toward them. Hence, much of what a person knows, values, and feels has a social origin (is shaped by others) and a social orientation (is directed toward others).

Even if the notion of "inner life" seems to refer to the thoughts and feelings of individual people, we must emphasize that "inner life" is a fundamentally social category. We cannot imagine any inner subjective life without the intersubjective dimension – and vice versa. Moreover, the concept of culture – collective ways of knowing and feeling – is also relevant here. We can even conceive of an inner life of collective categories such as families, groups, communities, and organizations. One might, for example, think about the concept of "group identity." A concept such as culture is therefore highly relevant, because it deals with shared assumptions, worldviews, and ways of understanding and perceiving that exist within a community or society. It has to do with our social stock of frames of reference, our ways of knowing, valuing, and feeling.[1]

Secondly, the social implies *relationships and interactions* with others, both close and distant. These interactions are guided by social roles/positions, habits and norms, beliefs, and the various everyday rituals we perform together to confirm social bonds, for instance family dinners, birthday parties, socializing in cafes, and so on. Language and other cultural systems of signs and communication facilitate such interactions (Goffman 1967/2005, Blumer 1969, Collins 2004). These interactions can be spontaneous or organized, planned or unplanned, direct (face-to-face) or mediated (e.g., through social media). Agency and meaning are often activated through our social relationships and interactions. Some of these interactions are more personal; they are part of our "in-groups." Others are filtered or formalized through social roles or positions in organizations and relationships of power. Indeed, social life is to a considerable extent constituted by formal and informal social relations and interactions. We can even say that social relations and interactions are what make human life meaningful. We cannot live without any social belonging, and therefore social validation by the groups we seek to belong to is extremely important. Furthermore, what we accept or reject as true or false is strongly shaped by such group membership; not only worldviews and attitudes are fundamentally social in nature, but also what we hold to be true and what we perceive as knowledge (Klintman 2019).

Social relationships and interactions are often seen as ends in themselves. We want to have good relationships because they validate us as human beings. Sometimes, however, social interaction is used as a means to other ends. For example, you might interact with someone because they have something that interests you, perhaps an idea, a resource, or a product. So you engage in an exchange with that person. Or you might interact with someone out of necessity, as a consequence of power relations and systems of authority. When it comes to considering social interaction as a means to another end, a very popular and seemingly peculiar concept is *social capital* (Coleman 1988, Putnam et al. 1993). This idea carries the implication that "the social" is a resource that can be used to realize aspirations of various kinds. It refers to an actor's network of existing or potential contacts (interactions) that can be utilized for various purposes. Unlike money, social capital is not consumed when it is used; on the contrary, it may even be strengthened when used.

Thirdly, the social is embedded in *socio-material entities and practices/habits*. These include entities such as our own bodies (think of plastic surgery), apps, homes, offices, landscapes, shopping malls, physical, technical, digital or administrative infrastructures, and all kinds of technologies and habits such as dining, commuting, and shopping. Accordingly, social life always takes place somewhere, at some particular time, with some equipment, and according to certain routines. This conception of the social is often stressed in theories of social habitus and social practices (Shove et al. 2012, Bourdieu 2019), in actor-network theory (Callon 1986, Latour 2005), and as part of the concept of material culture (Miller 2010). These theories emphasize how social life is intertwined with the immediate material surroundings, which include both technological and cultural elements.

Such socio-material entities and practices can structure both the spatial and temporal dimensions of everyday and social life. Environmental sociology is interested in studying the often-conflicting temporalities of ecological systems and social systems (Lockie and Wong 2018). The temporalities of social life (its rhythms, time scarcity, timeframes) do not always align with those of nature, often being focused on the demands of the present and neglecting long-term sustainability. Attending to socio-material entities – such as how the clock is inscribed in administrative and technological systems – and to their associated practices helps us to pay more attention to the forces that structure the temporalities of everyday life, for instance the demands for speed and synchronization (Rosa 2013, Sharma 2014, Southerton 2020). Think of the internet and ICT, and how they have facilitated rapid communication and transactions, social media, working from home, remote social relations, surveillance technology, home entertainment consumption, and much more.

Fourthly, the social refers to *stratification*. Society is stratified into different social classes (in the Marxist tradition) or status groups (in the Weberian tradition), or into other kinds of "us and them" categorizations. Different structural categories such as gender, ethnicity, class, and age also intersect, creating very different living conditions and life chances for people. Society is stratified in ways that put people in different positions with different resources, and this shapes how they act, feel, and think about various things. Theories of stratification draw attention to the problem of social inclusion, which is an important topic in the discussion of social sustainability. For example, contemporary societies face major challenges related to growing inequalities (see Chapter 3). While some people enjoy extreme wealth, others struggle to find shelter and put food on the table. Similarly, the global distribution of power resources is highly asymmetrical. Growing inequalities in regions, countries, and worldwide are much more than an economic problem. We will show how environmental problems and the broader environmental crisis closely intersect with the issue of inequality.

Finally, we need to draw attention to *institutions*, which create enduring, organized patterns of human life and activities. Institutions consist of positions, norms, and values within a particular structure (Giddens 1976, p. 96). They stabilize and structure action on a larger scale and with a high degree of permanence. The reason for this is that once institutions have emerged, they strongly condition and shape actions through socialization (Berger and Luckmann 1967/1991). Indeed, many terms used in the social sciences and in everyday language refer to institutions. The nation-state, democracy, the market, capitalism, and science are all examples of institutions, as are the family, work, media, and school. Institutions can exist at different levels of society, from local to global, and an institution – for example, the family – can have highly differing manifestations in different geographical and historical contexts.

Institutions organize actions and activities, thereby constraining and facilitating them. They shape ways of thinking, feeling, and acting; that is, they are internalized, part of people's inner lives (Berger and Luckmann 1967/1991). The fact that institutions constrain or facilitate action (both individual and organizational) does not mean that they are unchangeable. No institutions, not even nation-states and capitalism, are immutable. They change slowly and gradually, sometimes intentionally (through social struggle and political decisions) and sometimes unintentionally (as indirect consequences of human action). Occasionally they can change rapidly – as social or technological innovations, political struggles, or environmental disasters can make an institution obsolete – but usually they change gradually, and often in ways that we do not notice. Social science concepts such as "path dependency" and "inertia" are commonly used in institutional theory to refer to the stability and predictability of human life while at the same time stressing that, in the long run, even what seems solid and stable will change. The current call for social transformation means that certain institutions must undergo radical change.

We argue that the five facets of the social are deeply intertwined and interconnected. This means that any theory of social transformation must take into account all five facets; otherwise, it will be difficult to properly conceptualize and achieve

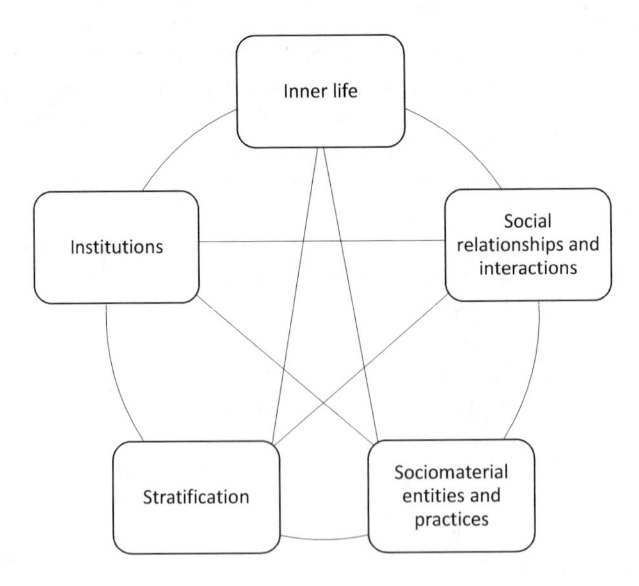

Figure 1.1 The five facets of the social

Source: Authors

social transformation. Including all five facets will also make it easier to predict the viability of policy measures and to prevent or mitigate unintended consequences of actions.

The word "social" also overlaps with other words, such as "society" and "community." Community often refers to something specific, concrete, and delimited, like the "professional community," the "urban community," or "the local community." We can also speak of an "imagined community." The concept of society may overlap with community, but it often refers to something larger or more abstract – for example a country ("Swedish society") as a broader entity than a nation-state ("the Swedish state"). "Global society" is the most abstract and all-encompassing level, referring to a global realm beyond individual states and countries (Sassen 2008, Beck 2009). Although society is a multifaceted concept with several different meanings, for our purposes it can be seen as the sum of all these social facets. Similarly, the treatment of "the social" within a particular community ought to take all five social facets into account.

At the most abstract level, our model of the social is presented as a universal feature of all societies. The five facets are constitutive parts of all human societies, but the contents of these parts are extremely variable in space and time. Indeed, societies and communities differ enormously. Thus, any analysis of the social must take into account the historical and spatial context.

Are we facing a civilizational crisis?

"We are in the fight of our lives," said UN Secretary-General António Guterres at the opening of the climate negotiations (COP27) in November 2022. Humanity, as he drastically put it, is "on a highway to climate hell with our foot – still – on the accelerator" (Guterres 2022a). The background to this sobering message is that while society has succeeded in solving some environmental problems, others are growing and now pose a threat not only to specific places and residents, but to humanity as a whole. The Intergovernmental Panel on Climate Change (IPCC) has found that climate change – through both slow processes like sea level rise and rapid ones like extreme weather events in the form of cloudbursts and storms – is drastically and progressively affecting our world's biodiversity and ecosystems (IPCC 2023). The Intergovernmental Science-Policy Platform on Biodiversity and Ecosystem Services (IPBES) has found that human actions are now threatening more species with global extinction than ever before (IPBES 2019). The depletion of biodiversity has recently been described as "the sixth mass extinction" (Kolbert 2014). Researchers have coined the term "planetary boundaries" to highlight that there are limits that humanity must not cross if we are to continue to develop and thrive. In 2009, the researchers estimated that humanity had transgressed three planetary boundaries: climate change, rate of biodiversity loss, and changes in the global nitrogen cycle (Rockström et al. 2009). In 2015, they found that a fourth boundary had been crossed, namely land-system change (Steffen et al. 2015), and seven years later, in 2022, the boundaries for novel entities and fresh water were also found to have been crossed (Persson et al. 2022).

Based on these findings published by scientists and expert organizations, we conclude that humanity is in great danger; it faces the risk of ecological and social collapse. There are even voices claiming that humanity needs to recognize the risk of its own extinction. Several books by journalists and columnists make this claim (e.g., Klein 2014, Scranton 2015, Thunberg 2022). In his award-winning book *The Uninhabitable Earth* (2019), David Wallace-Wells begins by saying "it is worse, much worse, than you think" (p. 3). These books are the latest additions to a large body of environmental alarmist literature, starting with Rachel Carson's book *Silent Spring* published a half century ago (1962), all of which stress that society is currently destroying its own environmental foundation. Some may see these voices as exaggerating the environmental threat, and think that this kind of environmental alarmism can lead to negative consequences – either to rapid and unreflecting action with unintended negative consequences, or to the opposite: pessimism, passivity, and cynicism.

At the same time, we believe this warning to be valid. If nothing is done, if society fails to change track, the habitability of the planet is at risk. The concept of the *Anthropocene* has been coined by the scientific community to capture this idea; humanity is not only impacting the environment, but also fundamental life processes on Earth (Lidskog and Waterton 2018). Human activities are now not only modifying ecosystems, but also transforming them, and thereby pushing the Earth system into a qualitatively different world. Global climate change is just the tip of the iceberg, because humans have changed not only the carbon cycle, but also other biogeochemical cycles that are fundamental to biological life (e.g., nitrogen, phosphorus, and sulfur) (Steffen et al. 2011, Williams et al. 2015). The challenges are summarized in what is called "the great acceleration" (Steffen et al. 2011). Diagrams show the dramatic increase in environmental problems (such as greenhouse gas emissions, ozone depletion, tropical deforestation, and woodland and biodiversity loss) and human activities (such as human population, GDP, damming of rivers, fertilizer consumption, water use and paper consumption). It is argued that the human enterprise "switched gears" in the 1950s (Steffen et al. 2011, p. 849) when the scale and rate of the human impact dramatically increased. Graphs depicting the great acceleration – using the "hockey stick" figure to illustrate the dramatic increase in consumption and emissions beginning around 1950 – have become iconic symbols of the Anthropocene (see Figure 4.1 in Chapter 4).

What concepts and expressions such as the Anthropocene, planetary boundaries, and the Great Acceleration point to is that the causes of environmental problems lie not in individual human activities, but in society as a whole (Lidskog et al. 2020). They also highlight that it is not the planet that is in danger, but society, humanity, and the biosphere. What is at stake is the living part of the environment, society with its living entities (both human and non-human). This is an anthropogenic, existential threat, a global threat to human civilization and survival caused by human actors (Dixson-Declève et al. 2022). We therefore believe it is essential to frame the current global environmental challenge as a civilizational crisis, a crisis of extreme proportions. If nothing is done, its ultimate consequence will be the extinction of the human race.

We also consider it necessary to frame the current challenge as a civilizational crisis because the roots of the crisis are deeply embedded in how present-day society functions. When we look for the roots of the current environmental crisis, we find that it is caused by our current civilization – what we believe and value, our ways of organizing and doing things, and our institutionalized ways of governing society. Therefore, we think it is not an exaggeration to speak of a civilizational crisis. In framing the current challenge as a civilizational crisis, however, it is important to do so in a way that does not conceal the diversity of the world or the societal progress that has taken place over the last century.

The framing of the global challenge as a civilizational crisis should not conceal the pluralistic and often contradictory character of world society. Globalization is generally regarded as a defining characteristic of contemporary society (Sassen 2006, Cohen and Kennedy 2013). In a very broad sense, it can be described as an interrelated set of technological, cultural, and economic innovations that extend our interactions with others across space and time (Ritzer 2011, Martell 2017). This means that we are all somehow implicated in the global environmental crisis, and we are all impacted in some way by the chemical and biological processes set in motion in other places, jurisdictions, and times. A similar concept to understanding the globe as a single place is to understand the global population as a single body of humanity, a humanity that is responsible for the current situation (Lidskog and Lockie 2020). Universalizing humanity in this way diverts attention from the socially unequal causes and effects of environmental problems – obscuring a socially stratified and polarized world, something to which we will devote a full chapter (Chapter 3). As postcolonial theory emphasizes, we need to constantly recognize and remind ourselves that environmental problems are shaped by a complex web of global, contextual, and historical factors in which various forms of domination and violence have played a significant role (Roos and Hunt 2010, Orlove et al. 2023). Any attempt to homogenize global society and environmental problems runs the risk of depoliticizing issues, concealing important aspects, and marginalizing or even silencing groups and communities. There is a great need for not only social transformation, but *just* social transformation.

On the other hand, in framing the global challenge as a civilizational crisis, one should not ignore all the societal progress of the past century. Global trends show astonishing reductions in infant mortality, child labor, and extreme poverty, as well as increases in life expectancy, literacy, democracy, and vaccination (Rosling et al. 2018). In short, for the majority of people, the world has become a better place to live. While many positive trends have recently been disrupted by wars and the Covid pandemic, remarkable changes have been achieved. Much of this development, however, has been made possible through exploitation of the Earth's finite resources and unequal global exchange (as we will discuss in more detail in Chapters 2 and 3). In some cases, the gap between what has been done and what still needs to be achieved is now widening. As UN Secretary-General António Guterres put it in his closing remarks at the COP27 climate negotiations (November 2022): "our planet is still in the emergency room … We can and must win this battle for our lives" (Guterres 2022b). The warning of the phrase "civilizational crisis" indicates both the severity of the current threats and the fact that we have not yet developed adequate means to handle them.

To speak of a civilizational crisis in the singular implies that the various crises facing society are interrelated and intertwined. Of course, many environmental problems have unique features and should not be lumped together. Climate change, for example, does not explain all the problems connected with biodiversity loss. It may even be the case that some of the means of mitigating climate change will result in decreased biodiversity – for example, large-scale planting of trees to create carbon sinks and/or to provide biofuels will lead to a monocultural forest system with substantial negative implications for biodiversity. Similar tensions and trade-offs can be found in most environmental areas. Furthermore, environmental problems are intertwined with many social problems, making it relevant to speak of a civilizational crisis in the singular. Of course, there are several social crises, each with its own characteristics, such as residential segregation, racism, and unemployment. Our reason for speaking of a civilizational crisis is to highlight that even if it is possible to understand and address these problems using different strategies and policy measures, they are intertwined with the global environmental challenge. For example, an unjust and highly segregated society is a major obstacle to initiating and promoting social transformation (see Chapter 3 for further discussion).

Ultimately, these crises may even threaten the survival of humanity as a species. They are "existential risks," that is, risks that are not only global but also threaten humanity as a whole (Boström 2013, Ord 2020). Even if humanity manages to cope with the civilizational crisis and thereby survives (as the authors of the book believe it can), we must not ignore the fact that the idea of human extinction is increasingly prominent in public discourse. Some people describe the imminent extinction of the human race as probable or even inevitable. Framing the global environmental challenge as a civilizational crisis is also a way to stress that, unlike with previous environmental challenges, it is not possible to solve this problem simply by prohibiting some activities, substituting certain substances or products, and/or modifying environmental regulations. The changes required are so deep and broad that the only adequate response is to restructure and reorganize society. The piecemeal and incremental change that some hope for will not suffice; what is needed is a fundamental and rapid social transformation. But what does social transformation mean, and what does it entail?

Social transformation

Both political bodies and scientific communities are beginning to call for society to be transformed to meet global environmental challenges (Díaz et al. 2018, Beck et al. 2021, Stoddard et al. 2021, Lidskog and Sundqvist 2022). Expert organizations such as the IPCC and IPBES stress that transformative change is necessary to meet environmental challenges (IPBES 2019, IPCC 2023). Environmental researchers argue that there is a need for transformative change to reorient society toward more sustainable pathways (Linnér and Wibeck 2019, Dixson-Declève et al. 2022). Political bodies consider transformative change to be a way to address environmental as well as social and economic challenges (see e.g., European Green Deal, EC 2019). Although transformative change is widely viewed as the right way forward and as an uncontroversial goal – it is hard to find anyone who is critical of it – its meaning remains remarkably unclear. It has become a buzzword in scientific and political discourses, where it is often sweepingly invoked as the solution to many serious environmental challenges. There is a dark side to its use as a concept (Blythe et al. 2018); the elites can adopt it, co-opt it, and use it to talk about sustainability without paying enough serious attention to real transformation.

As will be discussed in Chapter 4, there are different understandings and framings of the environmental challenge. Each of these frames opens and closes particular options, that is, what kinds of transformation are deemed relevant and justifiable. They also either depoliticize or politicize the environmental challenge, depending on the extent to which they challenge power relations and the social order. Some frames entail no need for deeper changes in society, while others suggest a need to restructure the prevailing economic and political power structures.

Thus, social (or societal) transformation is multifaceted and can be framed in different ways, leading to various proposals about what should be transformed and by what means. It is therefore meaningless to call for transformative change without specifying what is meant by transformation. We will elaborate on this in Chapter 6, but for now it is sufficient to follow Linnér and Wibeck (2019, p. 4) in defining societal transformation as "profound and enduring non-linear systematic changes, typically involving social, cultural, technological, political, economic, and/or environmental processes." Linnér and Wibeck say that it is easy to agree with such a definition, but assessing whether such change has occurred is another matter. Crucial questions include *what* should be transformed, *why* it should be transformed, and *by whom* and *how* it should be carried out. These questions are the subject of serious disagreement. Furthermore, by treating the social as constitutive (as opposed to the way the social appears in the quote above), social transformation must address all five facets of the social presented in the previous section. We will return to this in the concluding chapter.

A core message of this book is that in order to prevent a civilizational crisis, we must initiate and foster social transformation. Nothing is too big or too small to be changed, and all five facets of the social need to be involved. Thus, social transformation is not only deep, aiming to change the fundamental structures and institutions of society, but it is also broad, encompassing most parts of social life, from the inner life of people to global institutions. This is far from easy, not only because of the magnitude of the change required, but also because there are no well-trodden paths to follow. Instead, transformation demands ingenuity and social experimentation, as people and societies develop and test more sustainable ways of living and organizing society.

Thinking in terms of transformation is therefore a way of placing a problem in a broader context. Saying that transformative change is needed to address the climate crisis, for example, implies that what has been done so far is insufficient and ineffective. "More of the same" is not an adequate response. The political and scientific calls for transformative change signal a need for a deeper and more fundamental change. In striving for transformative change, any relevant and effective proposal must be based on a detailed and consistent view of how society is organized and how it is changing (Lidskog and Sundqvist 2022). Otherwise, it is very likely that the proposed solutions will work on paper but not in practice, or, if implemented, will lead to unintended consequences. Therefore, there is a need for expert analysis of the social causes of environmental problems. Why has a particular environmental problem arisen? Why does it persist, and how can it be changed? Without such analysis, there is a risk of inadequate solutions – solutions that claim to be appropriate but that may hinder more fundamental transformations. This will be further discussed in Chapter 5.

In seeking an informed way to understand society and how it changes, it is fruitful to begin with sociological thinking. Sociology has always been interested in social change. Indeed, much sociological thought has been developed to understand why social change occurs and what its implications are. In classical sociology, a key area of interest was the roots and consequences of major social transformations, such as those associated with industrialization and the growth of capitalism (Marx), urbanization (Simmel), social differentiation (Durkheim), and bureaucratization and rationalization (Weber). Underlying these theories of social transformation were the assumptions and worldviews inherited from the Enlightenment and the idea of unending progress, albeit accompanied by a host of social problems such as alienation, anomy, and the iron cage of the bureaucracy. Although the theories differ greatly, they all saw society as following the path of progress and "modernization." Modernization was first backed up by the institutions of science, and then by democratization and the growth of the welfare state.

Evidently, over the decades and centuries, this development and progress came to a halt in several ways. The fascism of the 1930s and 1940s served as an important reference point for sociological studies of modern society. Among the best-known of these are the works of the Frankfurt School and critical theory (see, e.g., Adorno and Horkheimer 1947/2016) as well as Zygmunt Bauman's (1989) influential work on modern bureaucracy as an institutional prerequisite for the Holocaust. The situation in many colonies, before and after they acquired independence and sovereignty, served as another important point of reference in the development of critical social science, not least in postcolonial studies (see, for example, the work of Said (1978) and Chakrabarty (2000), who show how power relations, discourses, and identities were maintained despite the abolition of colonialism). The growing awareness of the environmental problems that society was creating was also important. Environmentalism and environmental movements proliferated in several countries (O'Riordan 1981, Jamison et al. 1990). This spurred universities to develop environmental education programs and to invest more in environmental research. Some of this research stressed that environmental problems have a systemic character, that is, are linked to the way contemporary society functions and is organized. The well-known thesis on reflexive modernization, formulated by Ulrich Beck and Anthony Giddens, is indicative of a quest for a radicalized modernity. They agree that although the current version of modernity (which Beck calls "simple modernity") has significant shortcomings, it has also brought with it the possibility and necessity for a more self-reflective society in which actors and institutions can develop their capacity to anticipate and even prevent many of the undesirable side effects of industrial growth.

As Chapter 2 shows, different theories within environmental sociology locate the causes of the environmental crisis in different places, and accordingly propose different ways of transforming society to respond to the current environmental challenge. None of these approaches claims that social progress needs to stop, but they do claim that we need to rethink what we mean by progress, and that there are many different ways to progress.

The five thematic issues of the book

Causes: The social roots of environmental problems

The next chapter looks at the causes of our current environmental situation and argues that their roots are fundamentally social. This means that it is not enough to make sweeping statements that the causes are to be found in "culture," "society," or "humankind," but that we instead need to explore how contemporary societies function and how they have been shaped historically. Why is it that despite our increasing awareness and knowledge about environmentally detrimental practices – such as those that contribute to greenhouse gas emissions – society does not prohibit the activities that cause these problems? The chapter presents several arguments as to why "the social" can never be ignored when attempting to make a comprehensive and adequate investigation of the causes of environmental problems. The five facets of the social, which include explanations at the macro, meso, and micro levels, equip analysts to grasp today's environmental challenges, because they address the fact that there is no single cause of ecological destruction – such as capitalism, industrialism, or individualism – but many intertwined and interacting ones.

The chapter begins by reviewing four macrosocial theoretical traditions that seek to explain why current society is unsustainable, and why it has been so difficult to change course toward a more sustainable society. We then discuss why individuals – embedded in their social lives – do not live in a sustainable way, or why even environmentally very committed people find it hard to adopt sustainable lifestyles. We also explore the role of organizations, in particular how they link people and society together and interweave societal institutions and individual actions. Finally, we integrate the different strands of this argument using an explicit power perspective.

Distributions: The social spread of environmental problems

Although the environmental crisis is a truly global phenomenon, its consequences are unevenly distributed. Depending on who they are and where they live, people face very different environmental impacts, and this should not be overshadowed by the global character of the environmental challenges. There is an urgent need to bring the topic of uneven environmental effects to the forefront of environmental debates and analysis. Although this chapter focuses on one of the five social facets, stratification, it also reveals how intertwined this topic is with the other four facets.

The chapter addresses how differences between human groups and categories – such as class, gender, ethnicity, and their intersections – relate to environmental issues. Social stratification shapes life chances, material welfare, and access to natural resources as well as other important resources. It exposes people to different hazards and risks, and gives them different opportunities and capacities to deal with these problems. Differences in material life circumstances – power differentials and economic inequality – contribute to the reproduction of environmental problems. Such differences also affect how willing and able people are to contribute to collective problem solving. Social stratification and unequal exposure to environmental problems shape our different and sometimes highly polarized understandings of environmental problems. Inequality is therefore intimately related to the possibilities for transforming society toward sustainability.

We begin by discussing three of the most common social determinants, class, gender, and ethnicity, and how they are intertwined. We then turn our attention to the field of environmental justice, which not only highlights the most vulnerable groups and places that are most hard-hit by environmental problems, but also shows that these problems are often caused by other groups that are far removed from the affected places. Thereafter, we discuss how it can be possible for a wide range of affected groups to participate, or at least be represented, in decision-making. Finally, the chapter shows that issues of inequality and poverty are not only of great social concern, but also of crucial environmental importance. Inequalities can cause environmental problems and make it more difficult to find legitimate and viable solutions. Thus, any proposal or program for transforming society toward sustainability must address the issue of environmental justice; otherwise it will not be realistic and viable.

Understandings: The social sense-making of environmental problems

Not only are there many causes of the global environmental crisis and different consequences for different groups, but there are also many different understandings of the current environmental situation, its causes, and relevant ways to handle it. This chapter explores how people and organizations make sense of environmental problems. This is related to all the social facets, from the inner lives of people and communities (emotions, values, knowledge, identities), social relations and interactions (how belongingness and conversations shape understanding), social practices and technological devices (such as carbon accounting), how our sense-making relates to socially stratified experiences, and institutional dimensions (such as the role of science in society).

We begin by stressing that our understanding of environmental issues is a broad social phenomenon and not just a compilation of scientific facts. The widespread call to "listen to the science" is therefore not enough to guide action, because knowledge alone does not motivate people and organizations to act. Instead, it is the wider story that really matters – how environmental problems and challenges are framed, narrated, and linked to norms, values, and emotions. To highlight this fact, we introduce the concepts of frames and storylines, and present three common framings of how to solve environmental problems: the technological fix, the structural fix, and the cognitive fix. We then explore how and why scientific facts can be deployed to support stories of both environmental progress and environmental decay. Even if science in itself is not enough, it is crucial, and so we end the chapter with a discussion of knowledge resistance, as exemplified by the phenomenon of climate denialism.

This ties in with what we stated at the beginning of this chapter: that what matters is our understanding of the world. Our understanding of the world helps us to orient ourselves in it and guides our actions, including the demands we place on others to act, and who we blame if they fail to act. This is the reason why many powerful actors spend so much money on greenwashing their products; they are trying to avoid accusations and threats to their business activities (through regulation or consumer boycotts).

Barriers: Social resistance through inadequate solutions

Today, there is a plethora of proposed solutions to the current environmental crisis. The reason for this is that solutions do not exist apart from our fundamental understanding of environmental problems. Instead, how we understand environmental problems – their causes and consequences, including their distribution – determines how certain actions are deemed effective and relevant, and others irrelevant or ineffective. In this situation, there is a growing need to critically scrutinize all proposed, and even claimed, solutions. The preferred solutions of various different actors are often presented in a piecemeal, limited, inadequate, or even misleading way. Rather than overcoming barriers to transformative change they conform to the barriers, and may even become obstacles to transformative change themselves. It is important not only to develop activities that are sustainable, but also to stop activities that are environmentally harmful. Otherwise, there is a risk that sound environmental policies will be thwarted by other policies leading in unsustainable directions.

We argue that we should not approach problems with piecemeal solutions, and we should avoid wishful thinking. What we need to do is to build a critical-constructive capacity to discern when packages of solutions are potential components of a broader process of transformative change, and when they are mere talk or not conducive to deep transformative change. Critical questions to ask are: in what sense are they *really* solutions? For whom are they solutions? What are they meant to solve? Are they real solutions to our socio-ecological crisis, to our civilizational crisis? Or could this talk of solutions be deceptive and only serve to defend business as usual? Do they change existing power structures and detrimental practices substantially, or only modify them slightly?

This chapter explores why it is difficult to initiate more radical and far-reaching transformative change. In the first section, we highlight two common barriers to change: complexity and resistance to change. We then discuss how the proliferation of "solutions" risks becoming a barrier to change rather than a facilitator of transformative change. This is discussed in terms of piecemeal solutions and the "problem-solving" mindset. We then recommend ten critical questions that an analyst can ask when scrutinizing something that is claimed to be a solution. Finally, we discuss the need to be critical of inadequate solutions, while remaining open to solutions with transformative potential, which is the topic of the concluding chapter.

Transformation: Ways of changing society

Transforming society is far from easy. Human agency and social structures are dynamically related, and it is extremely difficult to anticipate the consequences of decisions and actions. Nevertheless, transformation is possible. History is replete with both failed and successful attempts to change society, among the most recent of which are the invention, institutionalization, and spread of human rights, democracy, and public education. Thus, in addressing the environmental crisis, we must avoid both the extreme position of determinism – that society is too complex and remote from human agency to be possible to change – and the extreme position of voluntarism – that people can easily change society. We advocate instead a realistic possibilism – that society can be changed, but not without great effort and arduous struggle. In the concluding chapter, we continue our discussion of what social transformation means, but more importantly, equipped with the insights from the previous chapters, we offer a new perspective on the question of how we can transform society. We apply relevant theoretical insights and draw on a variety of empirical lessons from existing attempts to create change in society and communities. We structure our discussion around our theoretical model of the five facets of the social. In doing so, we hope to demonstrate the necessity of considering "the social" and the relevance of using our theoretical model for understanding the conditions for social transformation.

Note

1 Berger and Luckmann (1967/1991) presented a useful theory of how the internal and external worlds interact, a theory that is widely considered a classic work in the sociology of knowledge. They focused on how our societal stock of knowledge and understanding of reality are patterned and reproduced through a dialectical process of internalization, externalization, and objectification.

References

Adorno, T.W. and Horkheimer, M. (2016) *Dialectic of enlightenment.* London: Verso Books [originally published 1947].

Bauman, Z. (1989) *Modernity and the Holocaust.* Cambridge: Polity.

Beck, S., Jasanoff, S., Stirling, A., and Polzin, C. (2021) 'The governance of sociotechnical transformations to sustainability', *Current Opinion in Environmental Sustainability*, 49, pp. 143–152. https://doi.org/10.1016/j.cosust.2021.04.010

Beck, U. (2009) *World at risk.* Cambridge, UK: Polity.
Berger, P.L. and Luckmann, T. (1991) *The social construction of reality: A treatise in the sociology of knowledge.* London: Penguin [originally published 1967].
Blumer, H. (1969) *Symbolic interactionism: Perspective and method.* Berkeley: University of California Press.
Blythe, J., Silver, J., Evans, L., Armitage, D., Bennet, N.J., Moore, M-L., Morrison, T.H., and Brown, K. (2018) 'The dark side of transformation: Latent risks in contemporary sustainability discourse', *Antipode*, 50(5), pp. 1206–1223. https://doi.org/10.1111/anti.12405
Boström, M. and Davidson, D. (2018) 'Conceptualizing environment-society relations', in M. Boström and D. Davidson (eds.) *Environment and society: Concepts and challenges.* Cham: Palgrave Macmillan, pp. 1–24.
Boström, N. (2013) 'Existential risk prevention as global priority', *Global Policy*, 4(1), pp. 15–31. https://doi.org/10.1111/1758-5899.12002
Bourdieu, P. (2019) *Habitus and field: General sociology volume 2.* Cambridge, UK: Polity Press.
Callon, M. (1986) 'Elements of a sociology of translation: Domestication of the scallops and the fishermen of St Brieuc Bay', in J. Law (ed.) *Power, action and belief: A new sociology of knowledge?* London: Routledge, pp. 196–233.
Carson, R. (2002) *Silent spring.* Fortieth anniversary edn. Boston: Mariner Books, Houghton Mifflin Harcourt [originally published 1962].
Chakrabarty, D. (2000) *Provincializing Europe: Postcolonial thought and historical difference.* Princeton, N.J.: Princeton University Press.
Cohen, R. and Kennedy, P. (2013) *Global sociology.* 3rd edn. Houndmills: Palgrave MacMillan.
Coleman, J.S. (1988) 'Social capital in the creation of human capital', *American Journal of Sociology*, 94 (supplement), pp. 95–120. https://doi.org/10.1086/228943
Collins, R. (2004) *Interaction ritual chains.* Princeton, N.J.: Princeton University Press.
Díaz, S., Pascual, U., Stenseke, M., Martín-López, B., Watson, R.T., Molnár, Z., Hill, R., Chan, K.M.A. et al. (2018) 'Assessing nature's contributions to people', *Science*, 359, pp. 270–272. https://doi.org/10.1126/science.aap8826
Dixson-Declève, S., Gaffney, O., Ghosh, J., Randers, J., Rockström, J., Stoknes, P.E. (2022) *Earth for all: a survival guide for humanity.* Gabriola Island, British Columbia: New Society Publishers.
EC (2019) *European Green Deal.* https://www.consilium.europa.eu/en/policies/green-deal/ [accessed 4 October 2023].
Giddens, A. (1976) *New rules of sociological method: A positive critique of interpretative sociologies.* London: Hutchinson.
Goffman, E. (2005) *Interaction ritual: Essays in face-to-face behavior.* New Brunswick, N.J.: Aldine Transaction [originally published 1967].
Guterres, A. (2022a) 'Secretary-General's remarks to high-level opening of COP27', video of his opening speech at COP27, 7 November 2022. https://www.un.org/sg/en/content/sg/speeches/2022-11-07/secretary-generals-remarks-high-level-opening-of-cop27 [accessed 21 November 2022].
Guterres, A. (2022b) 'Statement by the Secretary-General at the conclusion of COP27 in Sharm el-Sheikh', 19November2022. https://www.un.org/sg/en/content/sg/statement/2022-11-19/statement-the-secretary-general-the-conclusion-of-cop27%C2%A0-sharm-el-sheikh%C2%A0%C2%A0 [accessed 21 November 2022].
IPBES (2019) *Global assessment report on biodiversity and ecosystem services of the intergovernmental science-policy platform on biodiversity and ecosystem services*, E.S.

Brondizio, J. Settele, S. Díaz and H.T. (eds.) Ngo. Bonn: IPBES secretariat. https://doi.org/10.5281/zenodo.3831673

IPCC (2023) *AR6 synthesis report: Climate change 2023.* https://report.ipcc.ch/ar6syr/pdf/IPCC_AR6_SYR_LongerReport.pdf

Jamison, A., Eyerman, R., Cramer, J. & Læssøe, J. (1990) *The making of the new environmental consciousness: A comparative study of the environmental movements in Sweden, Denmark and the Netherlands.* Edinburgh: Edinburgh University Press.

Klein, N. (2014) *This changes everything: Capitalism vs. the climate*. London: Allen Lane.

Klintman, M. (2019) *Knowledge resistance: How we avoid insight from others*. Manchester: Manchester University Press.

Kolbert, E. (2014) *The sixth extinction: An unnatural history.* New York: Henry Holt and Co.

Latour, B. (2005) *Reassembling the social: An introduction to actor-network-theory.* Oxford: Oxford University Press.

Lidskog, R. and Lockie, S. (2020) 'Globalizing environmental sociology', in K. Legun, J. Keller, M. Bell and M. Carolan (eds.) *The Cambridge handbook of environmental sociology. Volume I.* Cambridge: Cambridge University Press, pp. 30–46.

Lidskog, R. and Sundqvist, G. (2022) 'Lost in transformation: The Paris agreement, the IPCC, and the quest for national transformative change', *Frontiers in Climate* 4: 906054. https://doi.org/10.3389/fclim.2022.906054

Lidskog, R. and Waterton, C. (2018) 'The Anthropocene: A narrative in the making', in M. Boström and D. Davidson (eds.) *Environment and society: Concepts and challenges.* Basingstoke: Palgrave, pp. 25–46.

Lidskog, R., Berg, M., Gustafsson, K., and Löfmarck, Erik. (2020) 'Cold science meets hot weather. Environmental threats, emotional messages and scientific storytelling', *Media and Communication*, 8(1), pp. 118–128. http://dx.doi.org/10.17645/mac.v8i1.2432

Linnér, B-O. and Wibeck, V. (2019) *Sustainability transformations. Agents and drivers across society.* Cambridge: Cambridge University Press.

Lockie, S. and Wong, C.M.L. (2018) 'Conflicting temporalities of social and environmental change?', in M. Boström and D. Davidson (eds.) *Environment and society: Concepts and challenges.* Cham: Palgrave Macmillan, pp. 327–350.

Martell, L. (2017) *The sociology of globalization.* Cambridge, UK: Polity.

Mead, G.H. (1972) *Mind, self, and society: From the standpoint of a social behaviorist.* Chicago: University of Chicago Press [originally published 1934].

Miller, D. (2010) *Stuff.* Cambridge: Polity Press.

Mills, C.W. (2000) *The sociological imagination.* Oxford: Oxford University Press [originally published 1959].

Ord, T. (2020) *The precipice: Existential risk and the future of humanity.* New York: Hachette Books.

O'Riordan, T. (1981) *Environmentalism.* 2nd rev. edn. London: Pion.

Orlove, B., Sherpa, P., Dawson, N., Adelekan, I., Alangui, W., Carmona, R., Coen, D., Nelson, M.K. et al. (2023) 'Placing diverse knowledge systems at the core of transformative climate research', *Ambio*, 52, pp. 1431–1447. https://doi.org/10.1007/s13280-023-01857-w

Persson, L., Carney Almroth, B.M., Collins, C.D., Cornell, S., de With, C.A., Diamond, M.L., Fantke, P. et al. (2022) 'Outside the safe operating space of the planetary boundary for novel entities', *Environmental Science & Technology*, 56, pp. 1510–1521. https://doi.org/10.1021/acs.est.1c04158

Putnam, R.D., Leonardi, R. and Nanetti, R.Y. (1993) *Making democracy work: Civic traditions in modern Italy.* Princeton, N.J.: Princeton University Press.
Ritzer, G. (2011) *Globalization: The essentials*. Chichester: Wiley-Blackwell.
Rockström, J., Steffen, W., Noone, K., Persson, Å., Chapin, FSIII, Lambin, E., Lenton, T.M. et al. (2009) 'Planetary boundaries: Exploring the safe operating space for humanity', *Ecology and Society*, 14(2), p. 32. http://www.ecologyandsociety.org/vol14/iss2/art32/
Roos, B. and Hunt, A. (eds.) (2010) *Postcolonial green: Environmental politics and world narratives*. Charlottesville: University of Virginia Press.
Rosa, H. (2013) *Social acceleration. A new theory of modernity.* New York: Columbia University Press.
Rosling, H., Rosling, O. and Rosling-Rönnlund, A. (2018) *Factfulness: Ten reasons we're wrong about the world – and why things are better than you think.* London: Sceptre.
Said, E.W. (1978) *Orientalism*. New York: Pantheon Book.
Sassen, S. (2006) *Sociology of globalization*. New York: WW Norton.
Sassen, S. (2008) *Territory, authority, rights: From medieval to global assemblages.* Princeton: Princeton University Press. https://doi.org/10.1515/9781400828593
Scranton, R. (2015) *Learning to die in the Anthropocene: Reflections on the end of a civilization*. San Francisco, CA: City Lights Books.
Sharma, S. (2014) *In the meantime. Temporality and cultural politics.* Duke University Press.
Shove, E., Pantzar, M. and Watson, M. (2012) *The dynamics of social practice: Everyday life and how it changes.* Thousand Oaks, CA: Sage
Southerton, D. (2020) *Time, consumption and the coordination of everyday life.* London: Palgrave Macmillan.
Steffen, W., Grinevald, J., Crutzen, P. and McNeill, J. (2011) 'The Anthropocene: Conceptual and historical perspectives', *Philosophical Transactions of the Royal Society A*, 369(1938), pp. 842–867. https://doi.org/10.1098/rsta.2010.0327
Steffen, W., Richardson, K., Rockström, J, Cornell, S.E., Fetzer, I., Bennett, E.M., Biggs, R., et al. (2015) 'Planetary boundaries: Guiding human development on a changing planet', *Science*, 347(6223). 1259855. https://doi.org/10.1126/science.1259855
Stoddard, I., Anderson, K., Capstick, S., Carton, W., Depledge, J., Facer, K., Gough, C., et al. (2021) 'Three decades of climate mitigation: Why haven't we bent the global emissions curve?', *Annual Review of Environment and Resources*, 46, pp. 653–689. https://doi.org/10.1146/annurev-environ-012220-011104
Thunberg, G. (2022) *The climate book.* London: Allen Lane.
Wallace-Wells, D. (2019) *The uninhabitable earth: A story of the future*. Penguin Books.
Williams, M., Zalasiewicz, J., Haff, P.K., Schwägerl, C., Barnosky, A.D. and Ellis, E.C. (2015) 'The Anthropocene biosphere', *The Anthropocene Review*, 2(3), pp. 196–219. https://doi.org/10.1177/2053019615591020

2 Causes

The social roots of environmental problems

You've been having headaches for a long time and visit a doctor. The doctor says your headaches are mainly caused by excessive worry and prescribes that you stop worrying. If you do that, the headaches will probably stop. You're also advised to occasionally take headache tablets if the pain gets too strong, but the most important thing is to stop worrying. You leave the doctor's office and sit at home with a diagnosis of the problem, but do not really know any practical way forward, what to do to stop the headaches.

As we can see, this is a poor diagnosis, as the doctor has neither analyzed your situation nor used medical knowledge to suggest a relevant solution. It is probably hard for you to just let go of your worries, as they are symptomatic of something, such as relationship problems or fear of becoming unemployed. The headaches may also be caused by high workload and constant stress. To be able to propose a relevant remedy to your health problem, the doctor must make a diagnosis based on the underlying causes of the symptoms (the headaches) and thereafter investigate how these causes can be eliminated or managed.

When environmental researchers diagnose and propose solutions to environmental problems, we all too often see a logic similar to the one above. We are confronted with imperatives such as: "humanity must change its relationship to nature," "society must stop using fossil energy," and "people must consume less." While these are simple and clear solutions, and are not wrong in themselves, they do not focus on root causes, and therefore have limited effect. They do not ask *why* humanity has a particular relationship with nature, *why* society uses so much fossil energy, or *why* people consume so much. Today, however, there is a growing awareness that many of the most serious environmental problems have not yet been handled and solved because their fundamental causes have not been addressed. That is one reason why many countries are promoting renewable energy at the same time as they are subsidizing non-renewable energy. Political initiatives to develop carbon-neutral transport modes are launched alongside approvals for the construction of new airports. Focusing solely on the symptoms of environmental problems, or on their superficial causes, will never solve the problems, because the root causes remain.

A starting point for this book – and for social science environmental research in general – is that environmental problems are societal problems; they are caused

DOI: 10.4324/9781032628189-2

by society and must be solved by society. This means that "the social" (see Chapter 1) should no longer be seen as just one aspect of environmental problems, as something that sometimes needs to be included in an environmental analysis and sometimes does not. If we are focusing on root causes, we must ask fundamental questions about why society, despite possessing so much scientific knowledge and public awareness, continues to engage in activities that generate environmental problems. Fundamental questions need to be raised about the design and structure of society, and key drivers need to be addressed, such as worldviews, relationships, norms, activities, and power.

Thus, the "social" can never just be something that is tacked on to an environmental analysis, as one category among others that may be important to consider (for example, when implementing solutions proposed by politicians or experts). Instead, it is at the heart of any environmental analysis aiming for transformative change: why has an environmentally destructive activity emerged? Why is it being perpetuated or even expanded, despite our understanding of its environmental impacts, and even in the face of a rather strong international mobilization, in which many powerful actors, including nation-states and transnational corporations, claim to be working to solve the problem?

This chapter argues that we cannot fully and adequately explain environmental problems – small or large, temporary or irreversible, local or global – without considering the role played by society. There is no single social root cause of all kinds of ecological destruction, such as capitalism or industrialism, but rather many interacting ones. We stress the five facets of the social, ranging from people's inner lives to large-scale institutions, which means that we need to use different theories and concepts focusing on different aspects of social life and society. Accordingly, in discussing the causes of environmental destruction, this chapter draws on a variety of theories.

We begin by presenting four general theories of why present-day society is unsustainable. These are macro-social theories; that is, they concern how society is structured and functions. We then show how society and people co-constitute each other, which implies that social analysis must consider how macro- and micro-social processes are intertwined. The conditions for people's lifestyles and social practices are influenced by both their immediate surroundings (e.g., family, colleagues, and friends) and broader structures (e.g., global processes affecting working life, as well as norms and expectations), and also, on the meso level, by the myriad of organizations and institutions that link people and society together. Before the conclusion, we introduce a power perspective, and the conclusion summarizes the chapter by relating the key insights to the five facets of the social.

Macrosocial explanations: What has caused the current environmental situation?

Environmental problems have multiple causes, ranging from individual choices to global institutions and processes. When you stand in the grocery store deliberating about what food to buy to make lunch, your choice is determined by many factors, from individual preferences and abilities (What kinds of food do I like? What dishes can I cook?), to social relationships (Which foods are good for my kids? How can I impress my girlfriend with a nice dinner?), practices of food provision and the socio-material infrastructure of logistics (How quickly can I get the refrigerated goods home?), the global food production enabled by supply chains and national and international trade regulations, economic assessments made by transnational companies (Why do they dump prices on meat?), and the marketing of particular food choices (How could I end up buying this fast food?). All facets of the social are involved. The micro and the macro are embedded in everyday life.

The following subsections mainly focus on a few important macrosocial theories that point to overall historical processes such as "modernization" and "globalization" and "colonialism/imperialism," as well as to various key institutions of society at large that have created the current environmental problems, and which need to be radically changed to solve the current environmental crisis. We refer to four different, but partly overlapping, research traditions: (1) the treadmill of production and degrowth; (2) risk society; (3) ecological modernization and transition theory; and (4) postcolonial theory and political ecology.

The treadmill of production and degrowth

The theory of the treadmill of production (Gould et al. 2008, Schnaiberg 1977, 1980, Schnaiberg et al. 2005) applies a political economy approach to social and environmental degradation. Based on Karl Marx's (1818–83) theory of capitalist society, it stresses how the constant (capitalist) search for economic growth (surplus value) leads to resource depletion and environmental degradation. The term "treadmill of production" was coined to describe the current institutional framework that serves as a means for increasing the production, distribution, and consumption of goods and services. According to this theory, economic growth is the main goal of modern (capitalist) society, and this in turn has had many secondary effects, not least environmental ones. Technology and science – which some hope will be part of the solution to the environmental crisis – are, from this perspective, mainly subordinated to this treadmill. They are used to improve its efficiency by increasing the extraction of natural resources, for example, through the mechanization and chemicalization of extraction processes ("ecological withdrawals"). This also leads to increased emission of pollutants ("ecological additions").

The state and the labor sector are also a part of this treadmill. It is in their interest to increase economic growth and thereby create more tax revenues, employment opportunities, and material wealth. Therefore, there are few initiatives to transform power structures and economic activities. Political and economic powers conceal their agendas, presenting themselves frontstage as committed to sustainable development, while backstage continuing their environmentally detrimental activities. Initiatives and proposals that would slow down the treadmill of production are frequently depicted as utopian and unrealistic, and as leading to severe (economic) consequences. Science is also often part of the treadmill, and is mainly used by exploiters and developers as a means to increase the production, distribution, and consumption of goods and services. Even if there are studies that produce knowledge about human impacts on the environment, such research has less access to resources and power because it is trying to slow down the treadmill. The treadmill of production theory, originally developed in the late 1970s and early 1980s, paints a rather bleak picture of the environmental situation, but offers hope that environmental research, environmental organizations, and public awareness can induce policymakers and society to change track.

The concept of "metabolic rift" describes a conflictual dynamic between society and nature (Foster 1999). It starts from the assumption that because humans depend on and are a part of nature, they need to engage in exchange with non-human nature to survive. This exchange – the metabolic cycle – must take place in a sustainable way, without threatening natural processes that make life and reproduction possible. Based on a Marxist understanding, the metabolic rift perspective stresses that the current (capitalist) society has a form of exchange that disturbs important natural processes, thereby creating a metabolic rift. Many current environmental problems can be understood as metabolic rifts caused by society. For instance, capitalist society is currently disrupting the global carbon cycle with its extensive use of fossil fuels, thereby releasing long-stored carbon into the atmosphere and causing a climate crisis (Clark and York 2005). Thus, environmental problems originate from the organization of society. And society, in turn, cannot survive if this rift is widening. In the long run, a socio-ecological metabolism must be developed that is capable of healing these rifts.

Some other concepts that have attracted much attention in recent years have clear similarities with the treadmill of production and focus on critiquing the economic growth paradigm and the expansionist tendencies of capitalism (on capitalism, see Box 2.1). As with the theory of the treadmill of production, the aim of the analysis is to reveal how societies and politics have become dependent on maintaining economic growth to ensure welfare, employment, and social stability. An increasingly popular concept, originally developed within ecological economics and now entering environmental social science, is degrowth (Paulson 2017, Kallis et al. 2018. Hickel 2019). It focuses on the need to downscale unnecessary economic activities in order to reduce material and energy use. The message of degrowth is that it is possible to reorganize society and live well under a different political-economic system that has a radically lower resource and energy throughput. Another similar concept is post-growth (Jackson 2017, 2021), which we discuss in Chapter 6.

Box 2.1 What is capitalism?

Capitalism is an economic system that focuses on profit making. A prerequisite for it is the institution of private property, including its legal protection. A historical prerequisite has been the concentration and accumulation of private property in the form of capital (money, properties, and other economic resources). In capitalism in its purest form, we can distinguish between owners of capital (capitalists), who can make investments in the means of production (natural resources, production facilities, technology), and the majority of people (workers), who lack this ability and therefore must sell their own labor to obtain income for their subsistence. Capitalism also relies on *market exchange* to acquire resources for production and to sell the products.

Capitalism is a dynamic economic system, which means that it is geared toward continuous expansion rather than a steady state. The opposite would be a non-growth economy focused only on satisfying needs. The dynamic nature of capitalism is due to both the profit motive and market competition. Because of market competition, firms cannot remain satisfied with their existing situation, but must remain competitive, control market shares, and reinvest part of their profits to maintain facilities and promote technological innovation. Therefore, there is a never-ending drive to reduce costs. This feature has been a constant focus of Marxist research, which has stimulated analyses of the alienating and contradictory nature of capitalism (workers, by being forced to sell their labor and having to accept low wages, are constrained, both as human beings and as consumers). The ecologically destructive nature of capitalism is of more recent date and is exemplified by various theories presented in this section. The debate continues as to whether capitalism can be tamed through regulation or whether it has to be replaced by another economic system. Another debate concerns how and to what extent market mechanisms can be used for the greening of societies, with market mechanisms and the profit motive leading to innovative technologies that outcompete existing polluting practices.

For further discussion of capitalism, see Brand and Wissen (2021), Jackson (2021), Swedberg (2003, pp. 56–65), and Wallerstein (2004).

Risk society

Ulrich Beck's (1992) *Risk Society: Towards a New Modernity* is one of the most influential works of macrosocial analysis in recent decades. The theory focuses on the social transformation from an industrial society that produces wealth to a risk society that produces risks and social hazards. Just as modernization in the nineteenth century dissolved the structures of feudal society and gave rise to industrial society, modernization is today dissolving industrial society, and a new society is coming into being. This emerging society is the "risk society," which is just as much a distinct social formation as the industrial society was. The risk society is distinctly different from the industrial society in that it focuses on the environmental question and the distribution of risks, instead of on the social question and the distribution of wealth. In both types of societies, risks are socialized; that is, they are perceived as a product of political decisions and human action. But unlike the risk society, the classical industrial society saw risks as manageable side effects of the production of wealth. These risks were legitimated partly with reference to the production of wealth and partly through society's development of precautionary and compensation systems (such as social insurance). The risks and hazards of the risk society are different from those of the industrial society, in that they are more widespread and serious. For the first time in history, there is a potential for global catastrophes caused by political decisions and human activities, not least activities involving chemical, genetic, and nuclear technologies. In the risk society, the relationship between wealth production and risk production is reversed. The production of wealth is now overshadowed by the production of risks. These risks are no longer localized in time and space, and consequently can no longer be seen as "latent side-effects" afflicting limited localities or groups.

Beck's theory of the risk society has some affinities with the treadmill of production. He links current environmental problems to the way central institutions of industrial society operate within what he calls "simple modernity." In simple modernity, problem solving depends on a science- and technology-centered approach in which uncertainty, complexity, and dilemmas are managed using instrumental rationality. Instrumental rationality depends on the view that it is possible to control reality through scientific knowledge and effective problem solving. Problem solving is specific and straightforward; the goal is to maximize efficiency and social and economic development (see also our discussion of the problem-solving mindset in Chapter 5). "Reflexive modernity" holds that the simple modernity approach to problem solving inevitably leads to unintended negative consequences. Science and technology have created false trust; the dogmas of scientific certainty and technological infallibility have produced widespread belief in the possibility to control all kinds of risk. Risk consciousness has been prevented from emerging (Beck 1995, p. 8). When side effects multiply, and traditional instrumental approaches are increasingly seen as irrelevant for handling them, a reflexive turn occurs. Beck places some hope in this development, in which actors will become more reflective and skeptical.[1] Somewhat ironically, when actors with diverse interests face global environmental disasters, they will start to find alternative visions of the common good. People will develop "new ways of seeing the world and being in the world," which in turn can create possibilities for global action (Beck 2015). Thus, enormous global risks also have a positive side, in that they provide new opportunities for emancipation from current ways of organizing society.

Ecological modernization and transition theory

In contrast to the theories of the treadmill of production and risk society, the theory of *ecological modernization* stresses that modernity is characterized not only by growing ecological degradation and crisis, but also by environmental reforms (Buttel 2003, Mol and Spaargaren 1993). Starting in northern Europe, and increasingly in other parts of the world, social, cultural, and political shifts have taken place, with industrial societies responding to environmental problems and taking measures to counteract them. With the help of scientific research, new (low-impact) technologies, reformed regulatory frameworks, and growing environmental consciousness, societies have begun to ecologically modernize themselves, replacing reactive, end-of-pipe solutions with preventive and proactive measures (Mol 1996). Not only policies but also practices have changed, and environmental performance has been boosted in many areas, such as the prevention of transboundary air pollution and CFC emissions. Ecological rationality is slowly catching up with economic rationality, leading to institutional innovations and gradual structural changes in which economic growth no longer has to lead to ecological disruption (Mol 2010).

Ecological modernization theory stresses the importance of environmental reforms and the incremental character of structural change. It does not make a sharp distinction between industrial and risk societies, and stresses that current environmental policies and practices have had substantial positive effects. Even if corporations and states occasionally block environmental reforms, modernization processes will continue to lead to eco-efficiency and environmental reforms. In this sense, we see a similarity to Beck's reflexive modernity, where modernity begins to reflect on its own consequences and develops actions in response. The essential difference is that while for Beck there is a tension and an ongoing struggle between simple modernity and reflexive modernity, ecological modernization theory sees reflexive modernity as a presently existing reality. However, change does not happen automatically. Reforms need to be backed up by nation states and supported by fruitful interactions with science and technology, markets, and civil society. One topic of study for this paradigm is to study why some countries – often in Western Europe – are more inclined than others to ecologically modernize their societies.

In its early stages, ecological modernization theory had a rather naive faith in science and technology, seeing it as a provider not only of truth but also of emancipatory power. This is close to technological determinism, the belief that technological development itself brings about social change and environmental reforms (see Chapter 4 on the technological fix). Over time, a more reflexive understanding has developed, in which science is no longer seen as consisting of undisputed facts, and its application as automatically leading to sustainable futures (Mol et al. 2014, Spaargaren 2000). A distinction is now made between large-scale technologies and softer or alternative technologies, and there is an interest in empirically exploring the role of science and technology, as well as other institutions such as the state, in environmental change. This is because globalization and reflexive modernization have challenged the institutions of science and technology (Mol and Spaargaren 2000). However, these institutions remain central to bringing about environmental reforms, not through technological fixes, but by making society more ecologically rational, that is, by institutionalizing environmental concern. Science and technology have an important, if not essential, role to play as progressive forces in the development of ecological rationality (Buttel 2000, Mol and Spaargaren 2000).

A similar framework is *transition theory* (see Geels 2011), which, like ecological modernization theory, focuses on long-term, technological macro changes. This theory argues that transitions in societies, such as shifting from one technological system to another (in transportation, energy, agri-food, or waste management) are generally complex, stepwise, and long-term, and involve a wide range of interacting actors and institutions (see also Chapter 6). Transition theorists adopt a multilevel perspective and use the concepts of niche, regime, and landscape. The term "niche" refers to entrepreneurship and bottom-up social and technological innovation. "Regime" denotes the established practices and "rules of the game" that stabilize the current system. "Landscape," the most macro-level concept, refers to external factors such as dominant ideologies, macro-economic conditions, and environmental factors that condition the regime. For example, the broad societal changes caused by the Covid-19 pandemic can be seen as landscape factors that have accelerated digitalization. Within this theoretical framework, Frank Geels and colleagues (Geels et al. 2015) promote a "reconfigurative" perspective in the study of transitions, which can be seen as something in between reform and revolution. They argue that research and policy should focus on the transformation of socio-technical systems and daily life practices in domains such as food, mobility, and energy provision. Socio-technical systems, in turn, consist of several elements such as markets, industrial structures, policies, consumption patterns, and cultural factors.

Both ecological modernization theorists and transition theorists have a fairly optimistic view of the possibilities for goal-driven management of society in the direction of sustainability. Transition theorists speak of "transition management" (Kemp et al. 2007) and argue for the importance of facilitating constructive interaction between key actors and systems at different levels. Creative niche actors must be given space for experimentation and upscaling. Obsolete regimes must be reformed. Governments have an important role to play as enablers or inhibitors. Some use the term "reflexive governance" (for an overview, see Boström et al. 2017, Leonard and Lidskog 2021), which refers to a self-critical and learning-oriented mode of governing through experimentation, feedback loops, and continual adaptation. Reflexivity is needed because of the inherently complex and uncertain nature of environmental problems and risks. Even if transition theory focuses more on pathways to effective change than on the root causes of environmental destruction, its adherents point to macro-factors that impede change. Indeed, regimes are stable by definition and seek to avoid disruptive change, so there is an inherent conflict between niche and regime. Factors preventing a transition toward sustainability include obsolete politics and "rules of the game" at the regime level, a lack of collaboration to facilitate upscaling of niche innovations, short-sighted lobbyism, vested business and political interests, rigid organizational structures, a lack of reflexivity, and technological lock-ins.

Postcolonial theory and political ecology

Postcolonial theory stresses that the current environmental situation – the social practices that have created it and now sustain it – are deeply shaped by histories of colonialism. Our culture (the way we understand and orient ourselves in the world), our epistemologies (the knowledge we create about the world), and our practices (the way we act in the world) are conditioned by our historical experience of colonialism. In this sense, colonialism is not a closed chapter in the development of the world, but persists and deeply affects what is happening today. Postcolonial research stresses that many theories of social change are implicitly embedded in a "colonial" way of understanding society, in which some crucial power relations are concealed (Bhambra 2014). This means that these theories are in fact ethnocentric, that they view and evaluate the world from a particular position without recognizing or acknowledging this fact. Macrosocial theories may therefore conceal social and environmental injustices caused by this postcolonial situation (see Chapter 3 for a discussion of environmental injustice and ecological violence).

For example, energy choices for decarbonization and possible energy futures might be analyzed without discussing how some options imply continued and deepened exploitation of already marginalized peoples and places (White and Roberts 2020). Similarly, e-waste recycling has led to extreme pollution in some urban areas, not least in African countries (Little 2021). Workers at e-waste recycling centers and scrap yards burn bundles of cables to extract copper, resulting in toxic exposure for the workers and the environmental contamination of places.

Closely related to postcolonial theory is *political ecology*. In response to "apolitical ecologies," which explain environmental conditions and injustices in terms of abstract processes such as modernization and demographic change, political ecology stresses that they are the result of fundamental political processes (Bryand and Bailey 1997, Perrault et al. 2015, Robbins 2020). It is the political and economic structures of the world that drive environmental and social change.

Postcolonial theory and political ecology have some affinities with the first group of macro-theories presented in this section, the treadmill of production, metabolism, and degrowth, because of their connections to Marxism and political economy approaches. But colonialists and imperialists are a central focus of attention here. For example, in their book *The Imperial Mode of Living: Everyday Life and the Ecological Crisis of Capitalism*, Ulrich Brand and Markus Wissen (2021) stress the importance of a postcolonial perspective and show how a global constellation of power and domination is reproduced at all levels of society and in a myriad of everyday activities.

The imperial mode of living, they maintain, is "deeply embedded in political institutions, the economy, culture and mentalities" (p. 39). For instance, they stress the powerful role of capitalism and neoliberal ideologies. "The concept of the 'imperial mode of living' points towards the norms of production, distribution and consumption built into the political, economic and cultural structures of everyday life for the populations of the global North" (p. 41). It is a mode of living based on inequality, power, and domination, and sometimes also violence, in relation to the global South.

This perspective stresses the importance of seeing and studying how the wealth of the global North depends on the repressive (and sometimes violent) extraction of natural resources and the production and distribution of cheap goods from the global South. This production is made possible by low wages and precarious working conditions, as well as low or non-existent environmental safety standards in the production context. This highly unequal pattern is also made invisible by a number of factors: neoliberal and racist ideologies, distance from the production context, as well as the way in which a sense of normality is produced by the mundane nature of everyday life in the wealthy context. This invisibility makes it difficult to challenge the existing norms of production and consumption. Brand and Wissen argue that this mode of living is fundamentally unsustainable; it can neither be extended to all populations on earth, nor to future generations. Thus, the internal contradictions within global capitalism will sooner or later necessitate transformative change on a global scale. Brand and Wissen do not place any hope in ecological modernization (including green capitalism, green growth, green consumerism) because the market-oriented and technological solutions of green capitalism only reproduce the same global power asymmetries that caused the problems in the first place.

All of the macro-oriented approaches mentioned in this section have a historical perspective. Contemporary institutions and infrastructures of society have evolved through previous developments (transformations, reforms, global power struggles, etc.). In Box 2.2, we briefly outline the main historical phases, drawing on the work of Brand and Wissen (2021). It should be noted that this sketch of the historical development is shaped by a particular macro-theoretical perspective. For example, other points would have been highlighted if we had used the ecological modernization paradigm.

Box 2.2 The historical development of the imperial mode of living

Brand and Wissen (2021) outline the historical development of the imperial mode of living of the global North in four broad phases (see also Lehning 2013). The main developments are summarized here in bullet points.

First phase (until the end of the eighteenth century): Early capitalism and colonization

- Breakdown of feudalism. Centralization of state power (mainly Spain, Portugal, Holland, Great Britain, and France are involved in colonization).
- Resource extraction based on land grabs and physical violence, including slavery.
- Ideological justification through racism and in terms of a civilizing mission in Africa and South America.
- Breakthrough of industrialism, facilitated by coal, the steam engine, and technological innovations.
- Urbanization and consequent concentration of workers in cities and large factories, facilitating the disciplining of workers.
- Growing bourgeoisie (upper and middle class) and taste for luxury consumption.

Second phase (nineteenth and early twentieth centuries): Liberal capitalism and advancing colonization

- Intensified global competition for labor power and natural resources.
- Further integration of racism in the imperial mode of living.
- Growth and consolidation of demand (consumer culture) among larger segments of society.
- Struggle for shorter working hours and better working conditions in the global North.

Third phase (world wars until the 1970s): Fordism

- Production of standardized commodities in large factories, facilitating "rational" mass production, markets, and consumption.
- Labor movement and class compromises between capital and labor: unions, shorter working hours, welfare.
- Struggle for a share of the abundance of commodities.
- Economic growth and increasing prosperity. Higher living standards.
- Growing critique of negative side-effects: environmental movements and alternative lifestyles.

Fourth phase (until today): Capitalist globalization

- Economic globalization: restructuring and deepening of the international division of labor; that is, a wave of outsourcing of domestic production to the global South (low wages, lax environmental standards).
- Neoliberal ideologies and justifications for this restructuring of the economy and politics, including austerity policies.
- Green capitalism and "green growth."
- Growing middle and upper classes in emerging economic regions (China, India, and Brazil) that adopt the imperial mode of living.
- Deepening of the imperial mode of living in the global North.
- Escalating planetary crises and deepening of unequal ecological exchange.
- Rise of authoritarianism and polarization.

Society as a driving force of environmental destruction

The macrosocial theories described in this section all share the view that environmental problems are unintended consequences. Environmental destruction is rarely sought after; only in exceptional cases (such as warfare and terrorism) does an actor intentionally aim to cause environmental destruction. All global environmental challenges, such as biodiversity loss, climate change, deforestation, and desertification, are unintended results (indirect effects) of human actions. The theory of the risk society is the clearest example of this. Until now, intended consequences have been legitimated partly with reference to the production of wealth and partly through society's development of systems for protection, precaution, and compensation (i.e., the positive and preventive sides of risk-taking). These risks and impacts are now becoming more widespread and severe than before, and it is therefore no longer possible to view them as manageable side effects, or as only affecting limited localities or groups. Economic growth, global trade relations, human planning, and technological development have brought not only material wealth, but also a dramatic increase in exploitation and risks, risks of such a magnitude that they are now transforming the fundamental structures of society.

The theories presented above share the view that there is a tension in contemporary society between the production of material wealth and the generation of environmental problems. However, they differ in their understanding of how current society works, which leads them to suggest different remedies. One way they differ has to do with the extent to which the fundamental structure of society needs to be transformed. Another difference concerns which actors will initiate and implement transformative changes and what means they will employ. Can states and corporate industries be the primary forces in combating environmental destruction? Or is it more important to raise public awareness, develop environmental research, and engage in social campaigns to pressure states to enact more radical environmental regulations? Can gradual environmental reforms (e.g., green taxes, smarter regulations, technological innovations) make society sustainable, or is there a need for more rapid and fundamental change? Might it even be necessary to establish a new economic system, as proponents of degrowth and some postcolonial theorists would argue? Will the way forward be fraught with violence, conflict, increasing polarization, and power struggles, or will different interest groups find ways to reach agreement about societal goals and the best ways to achieve them?

We will return to these issues later in this book, not least in the final two chapters on social barriers and social transformations (Chapters 5 and 6). For now, though, suffice it to say that a necessary task is to look for causes of environmental destruction at the macrosocial level (e.g., institutions). Another core message of this book is that environmental problems are caused by a composite set of factors, of which macro socio-technological infrastructure (part of facet 3) and institutions (facet 5), while important, are only two of many. Therefore, a comprehensive social analysis of environmental problems also needs to include the other social facets at the meso and micro levels, which will be discussed in the following sections and in later chapters.

Microsocial explanations: Why don't we live in an ecologically sustainable way?

A basic assumption of sociology is the duality of society. Society is maintained and transformed by our interactions and relationships, while at the same time it profoundly influences how we interact and relate to each other (Berger and Luckmann 1991). In other words, society and people are interdependent, inconceivable in isolation, and exert constant mutual influence. This duality means that human beings both shape and are shaped by society; they develop their thoughts, feelings, desires, and habits in particular social contexts, while also shaping and sometimes changing these contexts. The concept of social structure refers to how our actions and decisions are conditioned, while agency refers to the capacity of actors, whether individuals, groups, or organizations, to make decisions, take action, and change their circumstances.

Thus, environmental sociology cannot focus solely on macrosocial causes. In order to understand the causes of environmental destruction and suggest ways to work toward stopping this destruction, other aspects of the social must also be explored to reveal other kinds of causes that influence social life and social practices (Brewster and Puddephatt 2017, Čapek 2021). People's understandings and ways of acting in everyday life are therefore crucial.

What do we value and strive for?

The most obvious question to ask is that of why environmental problems exist. No one is seeking human extinction and planetary destruction. There are few, if any, anti-environmental values in society, in the sense of values that are oriented toward destroying the environment. So one might not expect there to be any obstacles to solving environmental problems. At the very least, there should be a willingness at the individual level to contribute to solving these problems. Nevertheless, we continue to engage in activities that are unsustainable, often despite our awareness of the risks.

There are several reasons for this. The first is that individuals, groups, and organizations have other, more immediate aspirations and goals than preserving the environment for the future, and efforts to achieve these goals involve exploiting natural resources and creating waste. Secondly, many of the environmental consequences of our actions are invisible to us. We do not experience them directly. Instead, they are reported to us by media and science (see Chapter 4). The environmental consequences are spread across time and space, occurring in other parts of the world or in the future. For example, current greenhouse gas emissions will cause future droughts in countries that are not major emitters. Therefore, these consequences appear abstract to us, and we cannot directly perceive the harm caused by our actions and decisions. Thirdly, we are socially embedded in relationships, groups, and organizations, and rarely reflect on the activities we perform in our different roles, especially when the pace of life is experienced as rapid and even accelerating (Rosa 2013, Southerton 2020). Fourthly, understanding how a particular action is linked to environmental destruction is an extremely complicated matter. Because it is difficult to figure out the most sustainable choice of T-shirt to buy and food to cook, people may feel they are not responsible for the environmental consequences of their actions. We have a case of "the problem of many hands" (Poel et al. 2015), where so many people are involved in creating a problem that their sense of personal responsibility fades away. This makes it easy to shift the responsibility for solving problems to other actors, and we tend to slip into free-riding behavior. The second, third, and fourth reasons all concern how easy it is for a person to feel powerless and helpless in the face of the great complexity of society (see Chapter 5 on complexity). The fifth reason, which is related to all of the above aspects, is that powerful interests influence our ways of living and acting. For example, enormous sums of money are spent on marketing to convince us, among other things, that some consumption is environmentally friendly, or at least is the most sustainable option. Through greenwashing – or greenblushing (hiding or not providing information about the environmental impact of a product or service) – we are persuaded to continue buying products and services in the belief that they are sustainable or climate neutral (see also Chapter 5). Taken together, these factors lead many of us, as individuals and groups, to continue living unsustainably and engaging in environmentally harmful activities.

Collective action could counteract these reasons why people cling to unsustainable practices (see Chapter 6). However, although environmental issues are on the political agenda, are widely reported in mass media, and are increasingly in the public consciousness, there is no strong and broad social mobilization to address them. Thus, not enough collective action is being taken, either by the "big players" in politics and business or by citizens and residents. Therefore, there is a need to better understand how we become and are human beings, which in turn will help us understand why people do not express their environmental concerns more loudly and take radical environmental action.

Becoming and being a person

Human beings are simultaneously individual and social; this is what is meant by the duality of society. Our inner lives are deeply intersubjective. We become individuals and develop a personal identity by interacting with others and looking at ourselves through their eyes, something that social psychology – beginning with the work of classical sociologists such as George Herbert Mead (1863–1931) and Charles Horton Cooley (1864–1929) – has studied in detail. This process does not end when we leave childhood and become adults. Throughout our lives we continue to view ourselves through the eyes of others and reflect on who we are and who we want to be. We continually modify ourselves and our self-understanding in relation to the people we interact with and the social contexts we are part of. Knowing who you are is very much about being aware of social contexts, adapting to them, and feeling a sense of belonging in them. You want to be like some people and different from others, and to belong to some groups and not to others. Belonging to a group often means that you are expected to behave in a certain way and even to possess attributes that indicate your group belonging. These are called identity markers, and they indicate which groups you belong to or want to belong to. External attributes that can serve as identity markers include clothing styles, language, behavior, and gadgets. The link between identity formation and (over)consumption is a core topic in the study of consumer behavior and (un)sustainable consumption, and is an important area to consider if we want to understand people's motivations for clinging to unsustainable consumer habits (see Box 2.3).

Box 2.3 Identity and (unsustainable over-) consumption

Given the growing recognition of problematic ecological footprints, climate footprints, and overshoot days, the relationship between overconsumption and climate change, along with other types of environmental destruction, has become a core topic in consumption studies. Scholars increasingly address the social and sociopsychological root causes of overconsumption (see Boström 2020, 2023), one important aspect of which is identity. For example, Helga Dittmar has spent decades researching the link between identity and excess consumption, highlighting the roles played by gender, advertising, and social peers. In a seminal article on consumption and identity, Russel Belk (1988) invented the concept of the "extended self" to describe the tendency to regard possessions as part of oneself. Objects that can be considered "mine" – even a person, such as "my wife" or "my child" – can easily be thought of as "me." Particularly during the "identity crisis" of adolescence, a young person enters into an identity-seeking process that involves acquiring and accumulating selected consumption objects, such as branded goods. Teenagers look to their peers to find and define themselves, and consumption choices are a crucial part of this (Dittmar 2008). However, consumer goods are important markers of identity at all stages of life, not least during major transitions such as getting married, having children, or advancing in one's career. Belk argues that it is especially during the middle years of life that people tend to have the most expansive conception of self. You develop your identity as a "respectable person" by owning a large house, a good car, nice furniture, and clothing with the right brands, and by traveling a lot. Similarly, as Zygmunt Bauman notes in his book *Consuming Life* (2007), modern consumer culture pushes everyone to construct themselves through consumption: "I shop therefore I am a sovereign individual, a subject" (Bauman 2007, p. 17).

Identity formation involves aspects of both inclusion and exclusion, which are two sides of the same coin. On the one hand, it relates to our basic human and social need to belong to groups, to fit in. We do not want to be excluded from our social group or to deviate from the norm, and we want to be seen as normal. On the other hand, it relates to an aspiration to stand out, to display status and distinction. Thus, there are both associative and dissociative reference groups. The sociology of consumption has a long tradition of researching and demonstrating how status and differentiation drive consumption (Boström 2023). Consumption should be understood as a signaling device that people use to position and differentiate themselves in the social landscape. Luxury consumption and profile brands can fulfill this role. Classical perspectives have been provided by sociologists such as Thorstein Veblen (1899/2009) and Pierre Bourdieu (1984/2010)

Sociologist Erving Goffman (1922–82) shows that our lives can largely be understood as theatrical performances (Goffman 1956/1990). Actors cannot do exactly what they want with their roles; the roles set some limits on what is possible. The actor must both meet the audience's expectations and play the role in a personal way. This, Goffman argues, is also what we do in our everyday lives. We want to make an impression on others, but we do not choose the role we play on our own. It is created in dialogue with our "audience," that is, those who observe us and on whom we want to make an impression. Goffman has been criticized for seeing all social interactions as role-playing, which implies that we are never honest in our self-presentation, but constantly think about the impression we are making. His response is that the roles we play actually show real sides of us. When we play a role, we do not do it in order to hide who we are. The roles we play are parts of who we are. We are, or will become, the characters we play.

We present ourselves to others through our appearance and behavior. The way we speak, move, and treat others tells people who we are. We act in ways that display our social status: behavior, body language, tone of voice, and emotional expression. We surround ourselves with things that tell others who we are: clothing, shoes, bags, cell phones, computers, bicycles, cars, and homes. Altogether, these things create a picture of who we are. Goffman describes in detail how we interact with others in different situations and present certain images of ourselves. The strategy we use is to try to make a certain impression on the spectators, what Goffman calls impression control. The image we want to give of ourselves is specific to the situation in which we perform the action. This means that we want the image to agree with the relevant context and the role we play in it. You act one way at a soccer game or a rock concert, but another way at a graduation or family celebration. This is because situations and groups have social norms for how to be and behave, and you adapt to these norms. Some colleagues at work might discuss the importance of combating climate change, and you might join in the discussion, stressing that individuals really need to consider their climate footprints. Back home, you and a friend then continue to plan a long-distance trip, perhaps entirely without reflecting on its climate impact. You adapt to the situation and speak in a certain way depending on who you are speaking with – relatives, friends, neighbors, colleagues, or people you do not know.

Presenting yourself through consumption

Goffman was not the first sociologist to study how we use different strategies to make an impression on others and to present who we are and where we belong. As early as the end of the nineteenth century, Thorstein Veblen (1857–1929), the founder of the sociology of consumption, stressed how consumption is used to display status, prestige, and power. The fact that some consumer goods can be effective in demonstrating social status was a key insight in his seminal work (Veblen 1899/2009). Veblen invented the famous concepts of "pecuniary emulation" and "conspicuous consumption." The former refers to the fact that much consumption has nothing to do with subsistence and everything to do with social status – the desire to equal or surpass the social status of others (which is demonstrated by exhibiting material wealth). The latter refers to extravagant and wasteful consumption, which Veblen exemplified with items that reflected the times in which he lived, including clothing, fashion, shoes, art, furniture, architecture, and rare animals. In contemporary times we can add high fashion brands, sports cars, boats, exotic travel, and multiple homes to the list of goods and services that people buy and invest in to display their wealth and high position in society.

Veblen argued that the restless pursuit of expensive goods emanates from the richest "leisure class." This tendency to engage in social comparison and status consumption then trickles down to the lowest strata of the social structure. The result is that everyone strives for more in an attempt to demonstrate status, but they all remain dissatisfied because the reference group, one rung higher on the social ladder, is always a bit richer. (This is somewhat like everyone in an audience standing on tiptoes to try to get a better view of the stage.)

Similar topics reappeared as key themes in Pierre Bourdieu's classic work *Distinction* (1984), where he explored how a "taste" for various consumer goods is related to social class and various forms of capital, such as cultural capital. We may think of taste as something personal, but Bourdieu emphasized the social nature of taste. Developing "good taste" often requires socialization and many years of unconscious training and cultivation. According to Bourdieu, like Goffman, the way we choose to present ourselves communicates our status in relation to others. Bourdieu emphasized that it is not only what we buy that matters, but also cultural knowledge about how to use these goods. He focused closely on art, but we engage in all kinds of impression management with various objects: housing, clothing, sports equipment, TVs, restaurants, vacation destinations, and so on (see also Boström 2023).

We may often engage in this kind of status comparison unconsciously. This is because the signals we use to try to impress our social surroundings have, to a considerable extent, been learned unconsciously through socialization. They are incorporated into our bodies as dispositions or habitus (Lamont and Lareau 1988). They vary across social classes and in different societies (see also Chapter 3), and when we observe the habits of members of other groups we may judge them to be inferior because we believe our own conduct is respectable, normal, and natural.

The social pressure to consume

In line with Goffman's and Bourdieu's theories of how social situations strongly influence our actions, researchers have pointed to what they call a fundamental attribution error, which refers to how people tend to overestimate the importance of fundamental character traits and underestimate the importance of the situation or context (Gladwell 2000, Heberlein 2012). People's tendency to underestimate the importance of context is one of the reasons why there is so much talk in the environmental debate about changing attitudes and disseminating information, despite the fact that research shows that behaviors and decisions are not simple reflections of attitudes, but are related to situational circumstances as well as much deeper processes of socialization and culture.

Of course, this should not be interpreted as meaning that we are either completely adaptable and behave entirely according to our social context, or that we are completely determined by social structures. During childhood, we are socialized to become an individual person, and we internalize social norms about how to behave and develop a sense of identity. Throughout our adult lives, we gradually change our identity, personal values, and outlook on life in a process called "secondary socialization." In most cases, these changes in identity and habitus take place slowly, and some parts of our identity are quite stable, despite our changing situations and experiences. Deeply held personal values and other aspects of our self-identity may lead us to adopt a particular stance even in situations and contexts where doing so may violate norms and adversely affect our relationships with others.

The fact that identity and habitus develop in interaction with other people means that identity is not just something you have, but also something you do. The contexts in which I participate and the people with whom I interact constantly and unconsciously affect me. It is not only what I and others think about me that shapes my identity, but also what I do and who I do it with. An identity is maintained and developed through social practices and patterns of action that we rarely reflect on. Thus, there is a reciprocal interplay between my self-image and the activities I engage in. My practices (how I live) express my ideas and perceptions of who I am and who I want to be, but at the same time these practices influence my self-image.

Identity thus encompasses social practices and emotional life; we not only think and act in certain ways, but also feel in certain ways. Many originally external social norms, held by the wider society or a group to which an individual belongs, have been internalized (unconsciously appropriated) and form part of our personal identity. External constraints can thereby become internalized demands. When our words or deeds violate a norm that we have internalized, we feel shame and guilt for failing to live up to the standards that we have made into personal values and duties. Even if we violate a group norm that we have not internalized, we may still feel shame because other group members show their disapproval; we have failed to live up to the demands and expectations of others. For example, studies have identified such pressures in the context of climate-impact consumption. The metaphor of a "glass floor" is used to point to sociocultural standards that people are forced to conform to, and that prevent them from achieving carbon footprint reduction goals (Cherrier et al. 2012). The glass floor refers to what is seen as the minimum level of socially required consumption. People who consume below this threshold transgress the socially dominant rules of consumer society and risk social exclusion. This is a topic that Zygmunt Bauman (2007) also discusses in his book *Consumer Life*. For example, maintaining your employability requires not only investing in a relevant education and developing a good CV. You may also have to display significant levels of consumption, the right tastes, and the knowledge needed to consume the right kinds of products in a socially correct way.

A feature of our modern society is that we are involved in many different contexts and groups where we develop different aspects of our identity. Our inner lives can sometimes feel messy, chaotic, and conflicted due to all these varying social situations and contexts, and our actions can be fragmented and contradictory as a result. These contexts and groups are also often profoundly shaped by consumer culture. We are frequently subjected to cross-pressure, with different groups and contexts placing conflicting demands on us, which means that we do not really know how we should be or what we should do. Or we may be forced to live our lives in very different ways in different settings, for example, at home with our family and at work. We are assigned different levels of status in different groups, and different sides of us are encouraged and others devalued or concealed. All groups both enable and constrain us at the same time; they encourage certain ways of thinking, acting, and feeling by discouraging other ways of doing these things.

Everyday life as socially embedded

Let us summarize how the microsocial perspective is necessary to gain a broader understanding of the causes of environmental problems, as well as of the problems themselves. How can we explain the fact that, although no one wants environmental destruction, we often act in ways that exacerbate environmental problems rather than solving them? This question requires a qualified answer that goes beyond simple, sweeping explanations, such as individual selfishness or irrationality. Rather, the microsocial perspective considers explanations in terms of how we all develop into people with personal identities, values, and life orientations.

A general answer to this fundamental question is that our lives consist of many more tasks, ambitions, and challenges than those related to living in an ecologically sustainable way, and through our socialization, social relationships, and group affiliations, we adhere to many more values and goals than just environmental ones. These social drivers are among the causes of ecologically unsustainable lifestyles in affluent contexts. They are perceived as immediate and compelling, whereas environmental considerations seem to be long-term and abstract. The focus is on the present rather than the future, especially when everyday life is experienced as fast and hectic, which is often the case today (Boström 2023). We are social beings who seek to be acknowledged by others. Even if people attempt to live in a more ecologically sustainable way, it can be difficult. As our discussion of the five facets of the social shows, people who adopt environmental values (inner life), may find that other social factors prevent them from realizing their ambitions, or pressure them to give their environmental values lower priority than other goals. The reason for this is that social relationships are a crucial part of being human, and some relationships are fundamental to our individual being and social belonging. While consumption is sometimes used to demonstrate identity and distinction, as discussed above, it can also be a way of expressing love, care, and friendship (Miller 1998). Many consumption-laden activities in everyday life, such as shopping, dining out, vacations, birthday parties, and concerts, are performed to develop and affirm social relationships and group solidarity (Boström 2021, 2023). Social relationships, intimate relationships within families, as well as relationships between friends and in one's wider social network, can all be sources of excessive consumption because people in such relationships affirm each other. They develop social bonds by sharing the pleasure that consumption can give. Many consumer items have highly valued symbolic meanings (Firat et al. 2013). Indeed, a failure to consume the right products in the right way and in the right quantities can be seen as a failure to take care of one's most important relationships.

There may also be somewhat different interpretations of what it means to live in a more ecologically sustainable way. Your expectations about what constitutes a good life, what needs are legitimate, and what kinds of consumption are reasonable are closely tied to social status and class belonging. If you are rich and claim to live in an environmentally friendly way, that might mean reducing your long-distance flights to a handful a year, continuing to drive a lot but switching to an electric car, and installing solar panels on your big house. This is a step in the right direction, but it is far from being an ecologically sustainable way to live. For someone with fewer resources, being ecologically sustainable may mean something completely different, such as going vegan, ceasing to buy clothes, and starting to bike to work. Because society is stratified, we have different social positions, resources, and roles depending on our social characteristics such as gender, class, ethnicity, and age (see Chapter 3). This means that not only do we have unequal resources and opportunities to realize our ambitions, but we also have very different ambitions.

Thus, our individual lives are always social. All five facets of the social – inner life, relationships and interactions, socio-material entities and social practices/habits, social stratification, and institutional patterns – are important in explaining the causes of environmental destruction. Mechanisms are at play here that motivate and legitimate the continuation and even acceleration of unsustainable practices (resistance to change is further discussed in Chapter 5). This emphasis on social embeddedness also brings the meso level into focus; our individual lives are organized, and even when we act individually, we do so as part of and within a broader social framework.

Meso-social explanations

In the above, we have stressed the duality of society – how both the macrosocial and microsocial levels are relevant to understanding how society works and how people behave. This is not to say that all aspects of society can be explained in terms of macrosocial structures and microsocial agency, that is, in terms of an external society imposing constraints on people and social action. Instead, it emphasizes that structure and agency exist at all levels of society, and that there are many more entities than just a general society and human beings. Meso-social explanations focus on how socio-material arrangements, organizations, and institutions are interwoven with the ways in which society functions and individuals act.

Everyday life as embedded in socio-material arrangements and organizations

Our lives are embedded in a world consisting not only of social relations but also of socio-material arrangements and organizations. Our interactions with other people are entangled with – enabled, facilitated, and sometimes even governed by – material objects, technical systems, and organizations. Socio-material entities help us to interact and communicate, and thus they make our social life possible. This socio-materiality not only provides opportunities for action and interaction, but also controls our interactions and makes us dependent on them. Four decades ago, cell phones, computers, and the internet were not part of everyday life, but today it is almost inconceivable to live without these technological devices and the infrastructures on which they depend. This means that we are largely ensnared by our socio-material context. Not using – renting, buying, or consuming – certain material things can have severe social consequences and, in the worst case, ruin our social lives. Our social practices develop in interaction not only with other people, but also with material objects and technologies. Many of our actions have the character of routines that we rarely reflect upon, and when we do, they seem difficult to change. As Deborah Lupton (2014) has shown in her book *Digital Sociology*, digital technologies have been incorporated not only into social institutions, but also into our understanding of ourselves, and our use of and exposure to digital devices – such as self-tracking devices and social media – have digitized the self and the body, and shifted the boundaries between private and public life.

Socio-material entities such as cell phones influence our everyday lives and social practices, but they do not do so as single entities that you can choose whether or not to use. Instead, they are part of larger socio-technical systems. In modern society, we have large-scale socio-technical systems for energy distribution, food provision, transportation, and communication, for example. The term "socio-technical system" was coined to stress that such systems consist not only of technical artifacts, but also of social organizations, knowledge, rules, and operators that make them work (Bijker et al. 2012). Thus, they are developed by society and constantly depend on society to function, while at the same time society is largely dependent on them. "Digital citizenship" is an example of this. As citizens we are largely forced to become digitalized, because society is highly digitalized (Hintz et al. 2019). Not only economic transactions, but also political actions – contacting a public authority, asking a politician questions – are carried out in digitalized forms. At the same time, researchers studying socio-technical systems emphasize that these systems are co-produced; they are maintained through our use of them, and we are dependent on them (MacKenzie and Wajcman 1999). They are changeable, but not easily, and they create expectations that are difficult to ignore or resist. The invention and proliferation of smart phones, for example, has given rise to new norms and expectations in social life, such as that people should always be reachable (Hobson 2019).

Our high and ever-increasing demand for water and energy can be viewed in a similar way to our growing reliance on digitalization. Elizabeth Shove, who is known for applying and developing social practice theory (see Box 2.4), shows in her book *Comfort, Cleanliness and Convenience: The Social Organization of Normality* (2003) how the historical "escalation of demand" for water and energy is related to household practices of comfort and cleaning (see also Rinkinen et al. 2021). She is eager to highlight the role of "inconspicuous consumption," that is, consumption not linked to the aspects of style, status, and symbolism emphasized by Goffman, Veblen, and Bourdieu. Various dynamics, including socio-technical arrangements and norms, are at play in the invisible normalization of habits of high consumption and resource use. Scientific practices, standardization, commercialization, and infrastructure facilities combine to legitimize new forms of consumption. For example, the standard that indoor temperatures should be stable at around 22 Celsius (72 Fahrenheit) all year round, regardless of geographic location, drives the building of specific socio-material facilities, creates expectations, and locks people into technological dependencies (heating and cooling technologies). Because certain indoor climate conditions have become "normal," these manifestations of comfort have become deeply embedded in the structure of everyday life in the domestic environment, and are part of everyone's expectations. Shove applies a similar line of reasoning to practices of cleanliness (the home, clothing, the body). Compared with indoor temperature, however, cleanliness is even more laden with symbolic and moral aspects, with conventions about clean vs. dirty, pleasure, personalization, social distinction, and freshness. The bottom line is that comfort and cleanliness are not simply expressions of personal preference. To understand how consumption patterns are changing, and energy and water demands are increasing – or could be reduced – it is necessary to consider all the facets of the social dimension.

Box 2.4 Sociological approaches to technology and material objects: Actor-Network Theory and Social Practice Theory

Actor-Network Theory (ANT), originally developed by sociologists Michel Callon, Bruno Latour, and John Law, stresses the relational aspects of action and criticizes the common distinction made between social, material, and technological aspects of reality. This means that common dualisms in social science need to be overcome, not least those between society and nature and between the social and the technological. According to this theory, dualisms do not explain anything, but rather need to be explained: how and why is a dualism created and maintained? A frequently cited example is Latour's (1999, p. 76) discussion of the two common standpoints in the debate over gun control: "guns kill people" and "guns don't kill people; people kill people." The first locates agency in technology and the second in people, and both positions, Latour argues, are deeply misleading, because it is a combination of them – the existence of an actor-network – that creates agency. Therefore, both human and non-human aspects (e.g. artifacts and things) must be included in an analysis, and studying action involves examining relational fields where connections and assemblages are continually being made. Common categories such as culture, science, technology, and politics should not be separated, but instead should be understood as co-constructed, and the task is to explain this co-construction.

In environmental studies, ANT has stressed the need to transcend the usual separation between nature and society. In his book *We Have Never*

Been Modern, Latour (1993) exemplifies this with the problem of depletion of the ozone layer. Instead of starting with an observed ozone depletion in nature that leads to an international response in society, one should start with practices. What practices are used to detect, measure, monitor, and regulate ozone layer depletion? Thus, it is not properties of nature or society that should be studied, but assemblages of people, technologies, activities, and institutions.

Social Practice Theory (SPT), derived from the work of sociologists such as Anthony Giddens and Pierre Bourdieu, and further developed by Theodore Schatzki and Andreas Reckwitz, has been applied to environmental sociology by scholars such as Elizabeth Shove (2010), Gert Spaargaren (2011), and Dale Southerton (2020). Like ANT, SPT attempts to overcome the split between structural and actor-oriented approaches in sociology with its central analytical focus on social practices. It expresses a view of the actor as routinized and socially embedded, as well as highly embedded in socio-material infrastructures and technologies. When applied to environmental sociology, this perspective suggests that it is more relevant to focus on the ecological footprints of practices such as cleaning, showering, commuting, and cooking than on the ecological footprints of individuals and their consumption choices. Following this approach, many behaviors related to sustainability issues are viewed as taking place at the intersection of material infrastructures (housing, transportation systems, food supply systems), temporal structures (work time requirements, the social need for synchronization, etc.), social norms (around cleanliness, what is appropriate to wear and eat, being digitally connected, etc.), and tacit knowledge (how to use household appliances, how to drive a car, etc.). Because people are highly dependent on existing social conventions as well as material, temporal, and technological arrangements in everyday life, they cannot easily change their practices even if they want to. Therefore, consumer information tends to be a weak tool for inducing change. Despite all this, SPT should not be seen as deterministic. Change is possible, but not without a more holistic and collective effort involving how material and technological aspects interact with all the other facets of the social, to speak in our terms.

We also highlight other theories that emphasize technology, such as some of the macrosocial theories presented in this chapter and in Chapter 4.

Meso-explanations highlight that organizations are not only mid-level structures, such as means by which a state influences citizens' actions, or citizens mobilize resources to promote a particular cause. They also have power in themselves, and function according to their own logics and objectives, as the next subsection discusses.

We act within and through organizations

Sociologists are often critical of the notion of personal decisions or individual acts. Instead, when individuals act, they are enabled and constrained by various social forces. They are part of larger collectives – families, organizations, professions, nations – all with their own goals, ideas, norms, and resources. While such collectives strengthen and empower individuals, they also imply constraint and control. Sociologist Göran Ahrne (1994) coined the term "organizational centaur" to express how we are part human and part organization. On the one hand, organizations are dependent on people. Even if organizations can be seen as collective actors, it is always (groups of) individuals who make decisions when the organization acts. On the other hand, an organization constrains individuals' behavior through the goals, norms, and resources it provides. Because we are part human and part organization, there is always a potential tension between our organizational and our human (personality, identity, convictions) sides. Individuals who act within an organization – be it a family, a volunteer organization, or a business – do so on behalf of both the organization and themselves. For example, when someone is recruited to a company and receives a mandate to represent environmental concerns within the organization, there is always a set of organizational goals, policies, rules, and norms that determine how and what that person can say on behalf of the environment. They may be academically trained in environmental science and have acquired knowledge and values from this education, but at the same time they are being paid to speak to the public on behalf of the organization (for dilemmas in representing organizations, see Boström et al. 2018). The environmental manager may have to respond to a variety of conflicting internal and external demands, but it is an uphill battle to oppose short-term sales and profit targets, especially when one's own job and salary are at risk. In addition, there is never any simple and direct way to "represent" the environment, because environmental issues are complex, ambiguous, abstract, contradictory, and often scientifically uncertain.

Organizations need to both encourage members' creativity (to develop) and constrain it (to achieve order). In the latter case, organizations use a variety of means to limit individuals' possibilities to deviate from organizational goals. They can use both hard measures, such as formal rules or incentive structures that influence career opportunities and pay raises. They can also use soft approaches to control individuals, such as attempting to strategically develop an organizational culture (see Box 2.5). Individuals may accept the need to compromise their beliefs because of the difficulty of finding another job. But compromises may also be accepted because the organization provides a platform for the environmental manager; it may give the manager the formal authority to act and speak for the environment, albeit with a somewhat limited mandate, and provide resources that empower the manager to do so. In the end, there is always the option of voting with one's feet and exiting the organization (Hirschman 1970/2004). Exiting is a way of drawing a boundary around what is reasonable and acceptable, and it constitutes an important power resource for the members of an organization (Ahrne 1994). Nevertheless, strategies such as whistleblowing and exiting are rarely used by members of organizations.

Box 2.5 Organization theory applied to environmental problems

Organizational structure. An important starting point for organization theory was Max Weber's study of bureaucracy and societal rationalization at the beginning of the twentieth century (Weber 1946, 1947). This began a long tradition of research on how different organizational configurations (e.g., centralized vs. decentralized, formal vs. informal, static vs flexible, simple vs. diversified, professional vs. mechanical) shape the operations of government agencies, corporations, and civil society organizations. Studies might ask questions such as: to what extent is organizational structure to blame for a lack of ability or flexibility to adapt to changing environmental conditions?

Neo-institutional theory. This theory generally focuses on how organizations adapt to institutions in society (legal, normative, cognitive) to seek legitimacy for their operations. The theory uses the concept of isomorphism to explain why organizations in a particular field tend to look similar in form, style, and reasoning. Much environmental research applies neo-institutional theory to the study of corporate "greening" processes, though often from a critical perspective, highlighting superficial "talk" and "hypocrisy" (greenwashing) rather than real organizational change.

Organizational culture refers to shared norms, rules, rituals, values, ideologies, and beliefs in organizations. We can speak of the "inner life" of an organization. These cultural factors can be formal or informal, stem from top-down or bottom-up processes, be known (corporate visions) or unknown (deeper assumptions and worldviews), and be hegemonic or divided into sub-cultures. Research on organizational culture can focus on how cultural orientations prevent or foster change processes, such as green technological innovations or corporate sustainability strategies. Organizational culture theory can also be used to explain unsuccessful change programs as being hampered by failures to address deeper cultural aspects of an organization or by the existence of competing cultural orientations. It can also be used to assess (a lack of) communication about environmental issues in an organization. Does the culture leave room for whistleblowers?

Stakeholder theory. Stakeholder theory was developed to consider the existence of interest groups other than shareholders that affect a company. Stakeholders are both affected by an organization and can affect its performance. The theory focuses on the most important concrete internal and external actors surrounding an organization: shareholders, employees, unions, customers, suppliers, environmental advocacy groups, local civil society, media, government agencies, etc. These stakeholders differ in terms of how much power, resources, and public legitimacy they have and how urgent their demands are. Stakeholder theory suggests that organizational managers will prioritize the demands of those actors they perceive to have the most power, legitimacy, and urgency. Thus, an environmental protection advocacy group may be seen as having less priority than, for instance, a labor union or a key supplier. Coalitions among stakeholders may be formed to increase their power.

Organizational risks and accidents. Charles Perrow (1999) coined the thought-provoking concept of *normal accidents*, which encapsulates the claim that very complex and tightly coupled technological systems have great, even inevitable, catastrophic potential. The problem arises in highly complex systems with high levels of interdependency between system components. Managers cannot have a complete overview, and a malfunction in one component can lead to a devastating chain of effects (nuclear accidents, rocket explosions, etc.). His provocative claim has sparked a scientific debate (see e.g. Hopkins 1999) about whether such accidents are inevitable and frequent, and whether it is possible to design more reliable systems. Research can investigate how technological, regulatory, and cultural systems interact. The question is whether it is possible to develop more reliable technological systems in combination with appropriate regulations, management strategies, and safety systems and cultures.

Some other important paradigms in organization theory, all of which may be relevant to environmental studies, include resource-dependence theory, network theory, and principal-agent theory. We invite the reader to explore the theories and key concepts briefly illustrated in this box. It should be mentioned that the theories and concepts are applicable not only to the study of the causes of environmental problems, but also to the study of efforts to understand them, seek solutions, and bring about transformative change.

Organizations as actors

Emphasizing the role of organizations in society requires paying attention to the fact that organizations themselves engage in action. It is not enough to say that organizations work through individuals, and that it is only individuals who act, even if their actions are facilitated by their organizations. What we want to emphasize here is that organizations are important social phenomena in their own right. This observation was made already in the early days of sociology. Émile Durkheim (1858–1917) stressed that social phenomena are "social facts." They have an existence outside the individual and are something to which people adapt, often unconsciously. Another classical sociologist, Max Weber (1864–1920), discussed different forms of organizations and found that modern organizations – what he called bureaucratic organizations – are highly specialized, with clearly defined positions and formal rules that lead their members to make decisions in similar and predictable ways. An organization thus functions as an autonomous entity in relation to the people who populate it. George Ritzer (2021), in his book *The McDonaldization of Society*, takes Weber's theory a step further. Through the guiding principles of efficiency, calculability, predictability, and control, you always know what you will get when you place an order at a burger place, regardless of who prepares the food. Ritzer claims that these principles have become the model for rationalization processes today. This means that organizations can act independently of their individual members.

Thus, an organization is much more than the sum of its members, because it develops its own goals, logics, and capacities for action. It can be seen as both an actor and a structure: an actor in the sense that it develops its own courses of action – we could say it has a collective "inner life" – and a structure in so far as it enables and limits the actions of others. Organizations thus exhibit what sociology calls the agency-structure duality; an organization is developed through human effort to serve human purposes, but once established, it often develops additional purposes and goals of its own and imposes external restrictions on its human members. For example, it may attract attention (such as from customers, investors, suppliers, and government agencies, in the case of a business organization), expand to encompass complex structures (achieving a division of labor and responsibility through vertical and horizontal differentiation), and eventually become a stable structure in society around which a large number of actors (individual and collective) orient themselves. A very large organization, such as a multinational corporation, despite being created by human beings, may appear unchangeable and uncontrollable in relation to the enormous sustainability challenges we face today. At the same time, historical experience shows that seemingly monolithic and enormously powerful organizations also change and, in the long run, even dissolve. It is important not to confuse inertia with immutability.

A question of power

Before concluding this chapter, we need to draw explicit attention to a phenomenon that cannot be neglected in a discussion of the causes of environmental problems: power. The notion of power is integral to many of the theories that we have discussed in this chapter, and it is sometimes treated implicitly, sometimes explicitly.

Power permeates all facets of the social. On the one hand, it belongs to the fourth facet of the social; it has a great deal to do with our stratified society, as we will see in the next chapter. On the other hand, power is a much larger phenomenon, something that energizes and shapes societies. From a broader perspective, power not only manifests itself through societal stratification (for example, the power that "capitalists" have over the "labor force" or, within a patriarchy, that men have over women), but also springs from our institutions and is both generated by and shapes socio-material arrangements (facet three), our social relationships (facet two), and our thinking and feeling (facet one). The power perspective thus becomes important for acquiring a holistic understanding of the social. In doing so, it is important to view power as something that can both hinder and enable social change. The efforts of social movements and action groups to change unjust situations and economic and political structures are often referred to as counter-power to stress that such groups mobilize power in order to influence powerful actors. We have seen this historically in the campaigns against slavery, in the struggles of colonized countries for national autonomy, and today in environmental movements' efforts to influence international environmental agendas and prevent powerful corporations and states from blocking the international work for sustainability. Power is often discussed in terms of "power over," which refers to the ability to coerce, dominate, and control, and "power to," which refers to the capacity to make something happen.

A useful framework for discussing aspects of "power over" is Steven Lukes's (1974) distinction between three dimensions of power. The first dimension is based on the classic idea (from social scientists like Max Weber and Robert Dahl) that actor A has power over actor B to the extent that A can get B to do something that B would not otherwise do. This is a straightforward definition of power (what Lukes calls "decision-making power"). Based on this definition, we can look at how one actor (such as a corporation) can exercise domination over others (such as a local community) through its ability to mobilize economic and other resources on which the other actor(s) depend for their subsistence. A drawback of this perspective is that it tends to pay attention only to observable resources, decisions, behaviors, and outcomes.

The second dimension of power pays attention to the many more subtle aspects of power that permeate political, organizational, and everyday life. This adds to the picture and draws attention to how actors can exercise power by keeping issues off the political agenda (what Lukes calls "non-decision-making power"). This involves exercising power to keep issues out of decision-making processes, for example by excluding questions from public discussion. It involves the ability to limit the scope of the political process to what is available for public consideration. Viewed from this angle, B cannot even express his concerns, because there is no setting in which B could do so. Mass media and civil society both play key roles here, with their ability to mobilize and exercise power to put issues on the public and political agenda.

Although the second dimension is useful for improving the analysis of power relations, Lukes argues that this kind of analysis is still too closely tied to the study of relatively observable behaviors, decisions, and conflicts. Both the first and second accounts focus too much on what resource-rich actors do or do not do. What is missing is the more "socially structured and culturally patterned behavior of groups, and practices of institutions" (Lukes 2005, p. 25). Accordingly, Lukes provides an important bridge to such sociological concepts as normalization, disciplinary power, manipulation, hegemonic ideologies, alienation, and so on. A may have power over B because A is able to shape B's preferences and control B's thoughts and desires, that is, B's inner life (this is what Lukes calls "ideological power"). For instance, in relation to sustainability issues, it is not difficult to see the importance of the role of the fashion and advertising industries and neo-liberal pro-growth ideologies in fostering an unsustainable consumer culture based on excess (Boström 2023). In local environmental conflicts, arguments are often made about how important some exploitation of nature is for securing jobs and local prosperity. People are persuaded, even manipulated, into believing that the existing circumstances are the best they can expect.

This brings institutional forces of power into focus, for example how multinational corporations are empowered by institutions (the fifth facet), such as capitalism and state power, and in turn exercise enormous power over societies, communities, and populations. Theorists such as Michel Foucault and Anthony Giddens have been interested in understanding how disciplinary structures and the modern state and other institutions have been able to extend the scope of governance into even the most intimate features of everyday life: our inner lives (facet one). They both wrote about the expansion of surveillance capacity in modern societies, thanks to the role of statistics, the accumulation of information resources, military power, and so on. They did this long before the breakthrough of the internet, long before the whistleblower Edward Snowden revealed in 2013 how deeply powerful actors can peer into the inner lives of citizens across the globe, and long before we began to wonder where artificial intelligence will take us in the future. Both Foucault and Giddens showed how social reproduction is grounded in the supposed "normality" and predictability of everyday routines. Similarly, and from a more Marxist perspective, the theory of the imperial mode of living discussed earlier (Brand and Wissen 2021) shows a similar type of interplay between the facets of the social. Later, we will highlight several other examples, such as the role of the "polluter elite" (Kenner 2019; see Chapter 3).

These perspectives on power are useful for understanding the themes of the book: Causes, Distributions, Understanding, Barriers, and Transformation. Regarding the last theme in particular, the reader should not forget that we must also acknowledge the positive notion of power: the power to act. This is the power to transform society along more sustainable lines. Both Foucault (2001) and Giddens (1987) recognized this. For Giddens, transformative capacity is "the ability to intervene in a given set of events so as in some way to alter them" (Giddens 1987, p. 7). This involves mobilizing resistance and counter-power as well as using a variety of material, cultural, social, and knowledge resources. We will return to this topic in the final chapter.

Conclusion

When exploring the causes behind the current environmental situation, there is a danger of falling into one of two extreme positions. One is to view the situation too deterministically, assuming that it is impossible for any actor – whether an individual person, an organization, or even a state – to change the current patterns. Society has trapped itself in an unsustainable way of functioning, and there are only limited possibilities to change the direction in which society is evolving. This view risks generating passivity and cynicism, and leaves no reason to try to understand how people think and act in the situation.

The other extreme is to have an overly voluntaristic view of human action, believing that the spread of environmental consciousness is enough to make society change track. In this view, even strong centers of economic and political power will cease their environmentally detrimental activities and develop sustainable practices, once they understand the urgency of doing so. This view is naive, and risks leading people to waste their efforts on ineffectual awareness-raising campaigns.

The sociological understanding that we present here is that social change is possible, but all changes are influenced by, and occur within, specific settings. Our use of the five facets of the social dimension is based on this understanding of social life and society. It is not enough to focus on just one facet; we must consider all of them to understand the causes of environmental destruction and why it has been so difficult to stop it. The macrosocial theories introduced in this chapter teach us about the broad institutional framework that has served as a means of increasing the material wealth of societies, but which is now leading societies down fundamentally unsustainable paths. These theories also reveal the enormous amount of work that has gone into promoting and legitimizing the delivery of prosperity to the wealthy regions and populations of the world. The institutional framework provides the context for the empowerment of huge organizations, which in turn can wield enormous power in societies and shape the conditions of life on Earth. While it may be tempting to attribute the causes of environmental destruction to macrosocial factors – like global capitalism and large multinational corporations – we need to remember that, from a sociological perspective, it is wrong to view society and individuals as distinct entities, and to understand society as a mere external environment for human beings, one that imposes constraints on their actions. Macrosocial processes can only be reproduced through the active involvement of people, and the search for the causes of environmental destruction must therefore also look to microsocial explanations. Moreover, as we will show in the final chapter, we cannot realistically perceive a transformative change unless people are involved and mobilized to participate in processes of change.

At the same time, it is equally wrong to view human beings as developing aspirations and wishes independently of their social conditions and settings. Society is something that also resides within human beings, in our inner lives. Our words and deeds are not merely individual acts; they are also social products, embedded in social relationships and interactions. Throughout life, people develop their identities, values, and ambitions, and this development always takes place in a social context. Powerful actors influence (manipulate) others to believe that existing circumstances are the best possible. Existing socio-material entities create technological dependencies. People may find themselves locked into the immediate physical and temporal circumstances in which they live and work, and therefore be unable to imagine that life could be different. As we will discuss in the following chapter, unequal and different social circumstances, as well as factors such as social class and gender, strongly influence the aspirations that an individual develops. Different social circumstances also provide different conditions for realizing these aspirations.

Questions for reflection and discussion

At the beginning of the chapter, we find the example of a doctor who suggests treating the symptoms of your headaches, not the causes. Can you give examples of political proposals that treat the symptoms but not the causes of an environmental problem?

Society consumes enormous amounts of fossil energy, and continues to do so, despite the fact that it is known to cause climate change. What do you think are the main reasons for this dependence on fossil fuels? Which of these causes are located at the macro, meso, and micro levels? How are they interrelated?

Can states and corporations be central forces in combating environmental destruction? Or is it more important to raise public awareness, develop environmental research, and engage in social movements to force states to enact more radical environmental regulations?

Can incremental environmental reforms (e.g., green taxes, smarter regulations, technological innovations) make society sustainable, or is there a need for more rapid and fundamental change? Might it even be necessary to create a new economic system, as argued by degrowth and some postcolonial theorists?

The chapter explains why we do not live in an ecologically sustainable way, and points to the importance of status and identity, in which consumption plays a central role. What do you think are the most important examples of consumption that demonstrate status and identity? How can status and identity be formed without engaging in environmentally harmful consumption? What obstacles do you think a person will face when trying to live in an environmentally sustainable way, and how can these be overcome?

The authors emphasize that our lives are embedded in a world that consists not only of social relations, but also of socio-material arrangements and organizations. What technological devices do you find necessary (and almost unthinkable not to have)? How do these devices affect your life? What are the advantages and disadvantages of using and depending on them? What positive or negative environmental impacts might these devices have?

Note

1 Beck (1994, pp. 5–8) argues that reflexivity (understood as society's need to handle its own unintended side effects such as environmental hazards) can lead to more reflection, but does not necessarily do so. For a discussion of reflexivity and reflection, see Boström et al. (2017).

References

Ahrne, G. (1994) *Social organizations: Interaction inside, outside and between organizations*. London: Sage.

Bauman, Z. (2007) *Consuming life*. Cambridge: Polity.

Beck, U. (1992) *Risk society: Towards a new modernity*. London: Sage.

Beck, U. (1994) 'The reinvention of politics: Towards a theory of reflexive modernization', in U. Beck, A. Giddens and S. Lash (eds.) *Reflexive modernization. Politics, tradition and aesthetics in the modern social order*. Cambridge: Polity Press, pp. 1–55.

Beck, U. (1995) *Ecological politics in an age of risk.* Cambridge: Polity.

Beck, U. (2015) 'Emancipatory catastrophism: What does it mean to climate change and risk Society?', *CurrentSociology*,63(1),pp.75–88.https://doi.org/10.1177/0011392114559951
Belk, R.W. (1988) 'Possessions and the extended self', *Journal of Consumer Research*, 15(2), pp. 139–168. https://doi.org/10.1086/209154
Berger, P.L. and Luckmann, T. (1991) *The social construction of reality: A treatise in the sociology of knowledge.* London: Penguin [originally published 1967].
Bhambra, G.K. (2014) *Connected sociologies.* London: Bloomsbury Academic.
Bijker, W.E., Hughes, T.P. and Pinch, T.J. (eds.) (2012) *The social construction of technological systems. New directions in the sociology and history of technology*. Cambridge, Massachusetts: MIT Press.
Boström, M. (2020) 'The social life of mass and excess consumption', *Environmental Sociology*, 6(3), pp. 268–278. https://doi.org/10.1080/23251042.2020.1755001
Boström, M. (2021) 'Social relations and everyday consumption rituals: Barriers or prerequisites for sustainability transformation?' *Frontiers in Sociology*, 6. doi.org/10.3389/fsoc.2021.723464
Boström, M. (2023) *The social life of unsustainable mass consumption.* Lanham: Lexington books.
Boström, M., Lidskog, R. and Uggla, Y. (2017) 'A reflexive look at reflexivity in environmental sociology', *Environmental Sociology*, 3(1), pp. 6–16. https://doi.org/10.1080/23251042.2016.1237336
Boström, M., Uggla, Y. and Hansson, V. (2018) 'Environmental representatives: Whom, what, and how are they representing?', *Journal of Environmental Policy and Planning*, 20(1), pp. 114–127. https://doi.org/10.1080/1523908X.2017.1332522
Bourdieu, P. (2010) *Distinction: A social critique of the judgement of taste.* London: Routledge [originally published 1984].
Brand, U. and Wissen, M. (2021) *The imperial mode of living: Everyday life and the ecological crisis of capitalism*. Verso Books.
Brewster, B.H and Puddephatt, A.J. (2017) *Microsociological perspectives for environmental sociology.* London and New York: Routledge.
Bryand, R.L. and Bailey, S. (1997) *Third world political ecology*. London: Routledge.
Buttel, F.H. (2000) 'Ecological modernization as social theory', *Geoforum*, 31(1), pp. 57–65. https://doi.org/10.1016/S0016-7185(99)00044-5
Buttel, F.H. (2003) 'Environmental sociology and the explanation of environmental reform', *Organization and Environment*, 16(3), pp. 306–344. https://doi.org/10.1177/1086026603256279
Čapek, S.M. (2021) 'The social construction of nature: Of computers, butterflies, dogs and trucks', in K.A. Gould and T.L. Lewis (eds.) *Twenty lessons in environmental sociology.* New York: Oxford University Press, pp. 13–27.
Cherrier, H., Szuba, M. and Özçağlar-Toulouse, N. (2012) 'Barriers to downward carbon emission: Exploring sustainable consumption in the face of the glass floor', *Journal of Marketing Management*, 28(3–4), pp. 397–419. https://doi.org/10.1080/0267257X.2012.658835
Clark, B. and York, R. (2005) 'Carbon metabolism: Global capitalism, climate change, and the biospheric rift', *Theory and Society*, 34(4), pp. 391–428. https://doi.org/10.1007/s11186-005-1993-4
Dittmar, H. (2008) *Consumer culture, identity and well-being. The search for the 'good life' and the 'body perfect'*. New York: Psychology Press.
Firat, A., Kutucuoglu, K.Y., Saltik, I.A. and Tuncel, O. (2013) 'Consumption, consumer culture and consumer society', *Journal of Community Positive Practices*, 13(1), pp. 182–203.

Foster, J.B. (1999) 'Marx's theory of metabolic rift: Classical foundations for environmental sociology', *American Journal of Sociology*, 105(2), pp. 366–405. https://doi.org/10.1086/210315

Foucault, M. (2001) *The essential works of Foucault, 1954–1984 Vol. 3 Power*. London: Penguin.

Geels, F.W. (2011) 'The multi-level perspective on sustainability transitions: Responses to seven criticisms', *Environmental Innovation and Societal Transitions*, 1(1), pp. 24–40. https://doi.org/10.1016/j.eist.2011.02.002

Geels, F.W., McMeekin, A., Mylan, J. and Southerton, D. (2015) 'A critical appraisal of sustainable consumption and production research: The reformist, revolutionary and reconfiguration positions', *Global Environmental Change*, 34, pp. 1–12. https://doi.org/10.1016/j.gloenvcha.2015.04.013

Giddens, A. (1987) *The nation-state and violence: Vol II of a contemporary critique of historical materialism*. Palgrave Macmillan.

Gladwell, M. (2000) *The tipping point*. Boston: Little, Brown.

Goffman, E. (1990) *The presentation of self in everyday life*. London: Penguin [originally published 1956].

Gould, K.A., Pellow, D.N. and Schnaiberg, A. (2008) *The treadmill of production: Injustice and unsustainability in a global economy.* Boulder, CO: Paradigm Press.

Heberlein, T.A. (2012) *Navigating environmental attitudes*. New York: Oxford University Press.

Hickel, J. (2019) 'Degrowth: A theory of radical abundance', *Real-World Economics Review*, 87 (19 March), pp. 54–68.

Hintz, A., Dencik, L. and Wahl-Jorgensen, K. (2019) *Digital citizenship in a datafied society.* Cambridge: Polity Press.

Hirschman, A.O. (2004) *Exit, voice, and loyalty: Responses to decline in firms, organizations, and states*. Cambridge, Mass.: Harvard University Press [originally published 1970].

Hobson, K. (2019) '"Small stories of closing loops": Social circularity and the everyday circular economy', *Climatic Change*, 163(1), pp. 99–116. https://doi.org/10.1007/s10584-019-02480-z

Hopkins, A. (1999) 'The limits of normal accident theory', *Safety Science*, 32(2–3), pp. 93–102.

Jackson, T. (2017) *Prosperity without growth: Foundations for the economy of tomorrow.* 2nd edn. London: Routledge.

Jackson, T. (2021) *Post growth: Life after capitalism.* Cambridge: Polity Press.

Kallis, G., Kostakis, V., Lange, S., Muraca, B., Paulson, S., and Schmelzer, M. (2018) 'Research on degrowth', *Annual Review of Environment and Resources*, 43(1), pp. 291–316. https://doi.org/10.1146/annurev-environ-102017-025941

Kemp, R., Loorbach, D. and Rotmans, J. (2007) 'Transition management as a model for managing processes of co-evolution towards sustainable development', *The International Journal of Sustainable Development & World Ecology*, 14(1), pp. 78–91. https://doi.org/10.1080/13504500709469709

Kenner, D. (2019) *Carbon inequality: The role of the richest in climate change.* London: Earthscan.

Lamont, M. and Lareau, A. (1988) 'Cultural capital: Allusions, gaps and glissandos in recent theoretical developments', *Sociological Theory*, 6(2), pp. 153–168. https://doi.org/10.2307/202113

Latour, B. (1993) *We have never been modern.* New York: Harvester Wheatsheaf.

Latour, B. (1999) *Pandora's hope: Essays on the reality of science studies.* London: Harvard University Press.

Lehning, J.R. (2013) *European colonialism since 1700.* Cambridge: Cambridge University Press.

Leonard, L. and Lidskog R. (2021) 'Industrial scientific expertise and civil society engagement: Reflexive scientisation in the South Durban Industrial Basin, South Africa', *Journal of Risk Research*, 24(9), pp. 1127–1140. https://doi.org/10.1080/13669877.2020.1805638

Little, P.C. (2021) *Burning matters: Life, labor, and e-waste pyropolitics in Ghana.* New York, NY: Oxford University Press.

Lukes, S. (1974) *Power: A radical view*. London: Macmillan.

Lukes, S. (2005) *Power: A radical view*. 2nd edn. Basingstoke: Palgrave Macmillan.

Lupton, D. (2014) *Digital sociology.* Abingdon: Routledge.

MacKenzie, D.A. and Wajcman, J. (eds.) (1999) *The social shaping of technology*. Buckingham: Open University Press.

Miller, D. (1998) *A theory of shopping.* Cambridge: Polity Press.

Mol, A. (1996) 'Ecological modernisation and institutional reflexivity: Environmental reform in the late modern age', *Environmental Politics*, 5(2), pp. 302–323. https://doi.org/10.1080/09644019608414266

Mol, A. (2010) 'Ecological modernization as a social theory of environmental reform', in M.R. Redclift and G. Woodgate (eds.) *The international handbook of environmental sociology*. 2nd edn. Cheltenham: Edward Elgar, pp. 63–76.

Mol, A. and Spaargaren, G. (1993) 'Environment, modernity and the risk society: The apocalyptic horizon of environmental reform', *International Sociology*, 8(4), pp. 431–459. https://doi.org/10.1177/026858093008004003

Mol, A. and Spaargaren, G. (2000) 'Ecological modernisation theory in debate: A review', *Environmental Politics*, 9(1), pp. 17–49. https://doi.org/10.1080/09644010008414511

Mol, A., Spaargaren, G. and Sonnenfeld, D. (2014) 'Ecological modernization theory: Taking stock, moving forward' in S. Lockie, D.A. Sonnenfeld and D.R. Fisher (eds.) *Routledge international handbook of social and environmental change*. New York: Routledge, pp. 15–30.

Paulson, S. (2017) 'Degrowth: Culture, power, and change', *Journal of Political Ecology*, 24(1), pp. 425–488. https://doi.org/10.2458/v24i1.20882

Perrault, T., Bridge, G. and McCarthy, J. (2015) *The Routledge handbook of political ecology*. London: Routledge.

Perrow, C. (1999) *Normal accidents: Living with high-risk technologies.* Rev. edn. Princeton, NJ: Princeton University Press.

Poel, I.V.D., Royakkers, L. and Zwart, S.D. (2015) *Moral responsibility and the problem of many hands*. New York: Routledge.

Rinkinen, J., Shove, E. and Marsden, G. (2021) *Conceptualizing demand. A distinctive approach to consumption and practice.* London: Routledge.

Ritzer, G. (2021) *The McDonaldization of society: Into the digital age.* 10th edn. Thousand Oaks, California: Sage.

Robbins, P. (2020) *Political ecology: A critical introduction.* 3rd edn. Hoboken, New Jersey: Wiley-Blackwell.

Rosa, H. (2013) *Social acceleration: A new theory of modernity.* New York: Columbia University Press.

Schnaiberg, A. (1977) 'Obstacles to environmental research by scientists and technologists: A social structural analysis', *Social Problems*, 24(5), pp. 500–520. https://doi.org/10.2307/800121

Schnaiberg, A. (1980) *Environment: From surplus to scarcity*. Oxford: Oxford University Press.

Schnaiberg, A., Pellow, D.N. and Weinberg, A. (2005) 'The treadmill of production and the environmental state' in M.R. Redclift and G. Woodgate (eds.) *New developments in environmental sociology.* Cheltenham: Edward Elgar, pp. 15–32.

Shove, E. (2003) *Comfort, cleanliness and convenience: The social organization of normality.* Oxford: Berg.

Shove, E. (2010) 'Beyond the ABC: Climate change policy and theories of social change', *Environment and Planning A: Economy and Space*, 42(6), pp. 1273–1285. https://doi.org/10.1068/a42282

Southerton, D. (2020) *Time, consumption and the coordination of everyday life.* London: Palgrave Macmillan.

Spaargaren, G. (2000) 'Ecological modernization theory and the changing discourse on environment and modernity', in G. Spaargaren, A. Mol and F. Buttel (eds.) *Environment and global modernity*. London: Sage, pp. 41–72.

Spaargaren, G. (2011) 'Theories of practices: Agency, technology, and culture. Exploring the relevance of practice theories for the governance of sustainable consumption practices in the new world-order', *Global Environmental Change*, 21(3), pp. 813–822. https://doi.org/10.1016/j.gloenvcha.2011.03.010

Swedberg, R. (2003) *Principles of economic sociology.* Princeton: Princeton University Press.

Veblen, T. (2009) *The theory of the leisure class.* Oxford: Oxford University Press [originally published 1899].

Wallerstein, I.M. (2004) *World-systems analysis: An introduction*. Durham: Duke University Press.

Weber, M. (1946) 'Bureaucracy', in H.H. Gerth and C. Wright Mills, *Essays in Sociology.* New York: Oxford University Press, pp. 135–178 [originally published 1922].

Weber, M. (1947) *The Theory of Social and Economic Organization.* University of Michigan: The Free Press [originally published 1914].

White, D. and Roberts, J.T. (2020) 'Post carbon transition futuring: For a reconstructive turn in the environmental social sciences?', in K. Legun et al. (eds.) *The Cambridge handbook of environmental sociology. Volume I.* Cambridge: Cambridge University Press, pp. 223–242.

3 Distributions

The social spread of environmental problems

"Poverty is hierarchic, smog is democratic" (Beck 1992, p. 36). Ulrich Beck's apt statement expresses the fact that modern environmental problems are democratic in the sense that no one can escape them. Thus, environmental problems can have an extremely broad mobilizing potential, threatening both economic and political elites, as well as ordinary citizens around the world. Rich and poor people could, in principle, unite in the struggle to change society in a sustainable direction. Global environmental problems concern everyone, because no one can completely escape a planetary crisis. Unlike many other social problems, environmental problems therefore have great potential to cross social, institutional, and geographic boundaries, and to mobilize broad social and geographic support. The global environmental challenge, as Beck points out, can then provide new opportunities for concerted action by placing all people – regardless of who they are and where they live – in a situation that calls for joint action. The risk society (see Chapter 2) creates risk positions that transcend class positions.

Nonetheless, Beck's claim has been criticized for underestimating the importance of social stratification and how unevenly people are affected by environmental problems (Curran 2013, Mythen 2005). Looking at the world, we see people struggling to find food and water for the day, while others are planning luxury vacations abroad. Some people are extremely well informed about climate change and try to minimize their climate footprint, while others know all about fashion trends and have shopping as their favorite leisure activity. Some people dream of getting a fast and expensive car without considering its fuel consumption. Others long for a society with well-developed and environmentally sustainable public transportation. It is in this world that environmental problems are created, a world that is divided and stratified.

The effects of environmental problems are also unevenly distributed. Although climate change affects everyone, its impacts are most severe for people who live in vulnerable places and have few options to protect themselves. Such people also have fewer resources to adapt to changing environmental conditions. Social stratification, with its attendant problems of inequality, disintegration, social tension, and class conflict, has been a central concern of sociology since its inception. It is well known that society is differentiated horizontally and vertically into different status groups, classes, and other types of groups. Yet these vast differences

DOI: 10.4324/9781032628189-3

are often ignored or only superficially understood in much of environmental research and debate. Many overarching concepts, such as sustainability, ecological modernization, resilience, and the Anthropocene tend to obscure this. Failure to recognize stratification not only implies a narrow understanding of environmental problems, but also that proposed solutions will have limited relevance and impact. Beck has also stressed this, not least in his later work (see, e.g., Beck 2016, Lidskog and Zinn 2024). The distribution of risks often runs along class lines, with wealth accumulating at the top of the social hierarchy at the same time as risks accumulate at the bottom. In this way, environmental problems reinforce the class divisions in society. However, the diffusion, globalization and severity of risks, at least in the long run, also block private (and costly) escape routes at the top.

Thus, the global environmental crisis concerns us all, but this should not blind us to how social stratification shapes the different life chances of various groups in society and the conditions for transforming society itself. Thus, this chapter demonstrates the importance of addressing the social distribution of environmental problems and their effects.

The chapter deals with how differences between groups of people belonging to various categories – such as gender, class, ethnicity, age, place of residence, income, education, and power – relate to environmental issues. Social stratification shapes life chances, material welfare, and access to natural resources, as well as other key resources. It makes people unequally vulnerable to hazards and risks. They have different opportunities and abilities to deal with problems that arise in their local environment. Differences in material life circumstances – economic inequalities – contribute to reproducing environmental problems. Such differences also affect people's willingness and ability to contribute to collective problem solving. These differences are important to take into account, from the local to the global scale.

While this chapter focuses mainly on one facet of the social, namely stratification, it is important to recognize how interdependent and intertwined they all are. Experiences of injustice and inequality (inner life) shape people's acceptance of and capacity for individual empowerment, the development of environmental consciousness, and collective action. Such experiences also affect the quality of social relationships and interactions (facet two), which are needed to collectively tackle environmental problems and to form environmental movements. Injustice and inequality are related to access to important socio-material entities and the capacity to engage in the various social practices (facet three) of everyday life. The environmental justice literature, as we will show in this chapter, demonstrates how the material surroundings (such as pollution and local exploitation of natural resources) shape material living conditions. Finally, social stratification is closely linked to the operation of overarching institutions and power structures such as capitalism, democracy, and the welfare state (facet five).

In this chapter, we will also show how inequality is deeply related to the possibilities for transformation, a topic that we will expand on in the concluding chapter. We start by presenting insights from environmental sociology regarding social stratification and key categories such as class, gender, and ethnicity, including the concept of intersectionality. We then explore the topic of environmental justice, which has become a broad and important field in the environmental sciences, and which shows how people suffer differently from environmental problems because of social stratification. This leads to the question of representation and participation: how can a wide range of affected groups be included, or at least represented, in decision-making? Finally, the chapter shows that issues of inequality and poverty are not only of great social concern, but also of crucial environmental importance. Not only are the environmental consequences of activities unequally distributed, but inequality can also cause environmental problems and make it more difficult to find legitimate and viable solutions and to mobilize action for transformative change.

Social stratification: Class, gender, ethnicity

Society is stratified according to social hierarchies and social distinctions. The classic case of social stratification in sociology is class, which sociology has studied since its inception. The topic stems from Marx's distinction between people who own the means of production (the capitalists) and the larger group of people who only possess their own labor. These larger groups of workers have no choice but to sell their own labor capacity to the capitalists. The system results in a mental state of alienation, as the capitalists extract surplus value (profit) from the labor in a competitive labor market. Later Marxist analysis has worked with a series of more elaborate distinctions along a vertical scale. People can occupy different, sometimes contradictory, class locations (Wright 1984) with varying degrees of ownership and control (of the means of production, other people's labor, and one's own labor and time) within an increasingly world-scale system (Wallerstein 2004). An important part of this theoretical tradition consists of theories about the possibility for a working-class movement to challenge this institutional structure and, through societal transformation or reform, gain (some degree of) ownership and control of the means of production.

Marxist thought has inspired various studies in environmental sociology, such as the theory of the treadmill of production (see Chapter 2), as well as work on economic democratization (Albert 2020, Löwy 2020). It has also provided important critiques of proposed solutions that lack a class perspective. A recent example of this is Matthew Huber's (2022) *Climate Change as Class War*, which forcefully argues that the climate movement has failed to address class, labor, and production, and that any successful strategy for decarbonizing society globally needs both to put fossil capitalism at the fore and to appeal to and mobilize the majority of the world's population, the working class. Another research stream extends Marx's theory of alienation to discuss how the public, who are dependent on the capitalist mode of production and an increasingly commodified material environment, become distanced and alienated from nature (e.g., Dickens 1996). A feature of all kinds of Marxist class analysis is that structural power is the central aspect of stratification. This means that even people with good wages – for example workers in coal mines or on oil rigs in rich countries – have very limited power over their labor, and thus belong to the lower class.

Max Weber agreed that ownership and control of the means of production is an important aspect of the class analysis. To this, he added a horizontal scale, introducing the concept of "status groups." These are groups that share similar (religious, cultural) ideas, aspirations, and ways of life. These ideas need not be as directly related to material circumstances as the Marxist analysis requires. In environmental sociology, this tradition has stimulated studies of how lifestyle segments and environmental values intersect with consumption patterns, for example. Such studies often use concepts such as "cultural capital," borrowed from the French sociologist Pierre Bourdieu. Scholars speak of an "eco-habitus" that develops mainly in middle-class groups (Carfagna et al. 2014), implying that, for some groups, high cultural capital and social status are associated with the adoption of ecological lifestyles, although these lifestyles do not necessarily correlate very closely with lower ecological and climate footprints.

Are women more ecologically conscious than men?

Society is accordingly differentiated both vertically and horizontally, but it is not only social class that is considered important for sociological analysis. Other important categories that significantly affect the opportunities and obstacles people face include gender, ethnic and racial categories[1], income, age, education, occupation, place of residence (rural/suburban/urban), sexuality, disability/ableness, citizenship/statelessness, and often also characteristics such as language skills. Looking at the topic globally, your country of birth has a huge impact on your life chances. It matters a great deal whether you are born in a developing or a developed country. All these different categories, positions, or locations in society shape people's diverse life chances and experiences. People may be more or less vulnerable to risks depending on which of the categories they "belong" to. It is therefore important to study how such differences influence the ways in which people are affected by environmental risks and problems, and how people respond to risks and problems with different ways of understanding (see Chapter 4) and acting. Environmental studies must not neglect such categories.

Take gender, for example. Why do women seem to be more willing and able to take responsibility for protecting the environment? It is well documented that women are more likely than men to exhibit pro-environmental attitudes and behaviors, such as driving less or engaging in politically or ethically motivated consumption, that is, considering environmental and ethical issues in purchasing decisions (see, e.g., Gundelach and Kalte 2021, Kennedy and Kmec 2018, Micheletti 2004). These findings are nearly universal and are independent of income level. Women with high incomes outperform men with high incomes, and women with low incomes outperform men with low incomes. A student of the subject will find a wealth of research confirming this picture. It may seem easy to conclude that women are simply more caring than men, that they have a natural tendency to be more environmentally conscious, while men destroy nature. The sociologist, however, avoids such biology-based reasoning. Rather, as Kennedy and Kmec (2018) argue, we need to look at the issue in terms of structural inequalities. By seeing it as a matter of social stratification, sociologists relate these gender differences to other sources of inequality, as well as to gendered ideologies, norms, stereotypes, and identities. These norms and identities shape expectations about whether and how people harm, protect, and care for nature in the different spheres of life, in political decision-making, at work, and in the home.

Structural factors – such as "the proportion of women in the labor force, the proportion of female elected representatives, average pay for women versus men, or the extent to which government policies incentivize men to engage in care work" (Kennedy and Kmec 2018, p. 306) – may be very important factors in explaining caretaking. Other factors that may intersect with environmentally harmful or benign practices include enrollment in higher education and access to health care. Thus, larger economic, political, and social structures influence how men and women are affected by the environment and shape their willingness and ability to care for nature.

Masculinity norms and how they relate to social stratification are the other side of the coin. For example, studies in the USA and Norway have found that climate denial attitudes are significantly associated with conservative white men (Krange et al. 2019, McCright and Dunlap 2011). Such attitudes are particularly prevalent among powerful, elite white males, who seem to view their wealth, pride, and inherited positions of privilege as under attack by the climate change message. Climate change denial provides them with an "identity-protective cognition" (McCright and Dunlap 2011, p. 1164). Not only do messages that climate change is a hoax serve their material interests, but such messages help to preserve some congruence in their inner lives (facet one) so that they can continue to enjoy their wealth and define it as well deserved and justified. Climate change denial is furthermore linked to the "masculinity of industrial modernity," which stresses the rationality of dominating nature, instrumentality, and economic growth (Anshelm and Hultman 2014). A study in the Norwegian context confirms that conservative white men are more likely to endorse these attitudes, and contributes empirical evidence that this is not just an elite phenomenon (Krange et al. 2019). Climate change denial is also related to xenosceptical nationalist attitudes, especially negative attitudes toward migration and immigrants.

Indeed, climate change denial seems to be merging with right-wing nationalism. There is a category of men, in both the upper and lower social classes, whose self-esteem is tied to their own or their nation's perceived great achievements. They tend to accept, justify, and defend the status quo. If they belong to lower income categories, they may still feel a sense of pride in what they perceive as superior ethnic and national affiliations. They may not see themselves, or be seen by others, as successful in life, but at least they belong to the right social group. They also form concrete social networks, socialize with each other, and maintain their social relationships (facet two). Through these networks and social interactions, they cultivate their social capital and establish a "status group." These networks help them confirm that their assessments are indeed correct. They may also be reluctant to deviate from the norms of the group, as this could feel like a threat to their entire social life. Social media algorithms facilitate this important process of self-confirmation and justification.

Climate change is thus seen as part of a larger complex of issues that all seem to threaten existing privileges, including some masculine identities, norms, and lifestyles. No one else should interfere with one's perceived natural right to run one's own business or to drive one's own motor vehicle. Again, we should not naturalize such gender norms and conclude that there is something universal and biological at play here. After all, not all men hold these values. Rather, it is people who happen to have been socialized and integrated into a privileged patriarchal position, within a particular structure of social stratification, who are more likely to adopt such attitudes. It is a matter of probability, not determinism.

Intersectionality: The intertwining of diverse social categories

The concept of *intersectionality* has become increasingly important in sociology over the last few decades, and more recently in environmental sociology as well. But what is it? The concept is already implicit in the discussion above. Intersectionality is concerned with how social identities such as class, gender, age, religion, ethnicity, and sexuality *intersect* at different levels of society, including at the level of individual experience. These categories reflect interlocking systems of privilege and disadvantage. According to intersectionality theory, it is an oversimplification to speak of power and vulnerability in terms of a single category, such as gender, because experiences and life chances are shaped by highly varied combinations of positions and social situations. "Male" and "white," as discussed above, is one such intersection, an intersection that correlates with climate denialism. Furthermore, experiences of sexism and racism are often interconnected. A wealthy white woman may experience patriarchal relations differently than a working-class black woman. Concepts such as these have been recognized in recent studies in environmental sociology (Ergas et al. 2021, Hoover 2018, Malin and Ryder 2018, Olofsson et al. 2016). For example, some specific intersecting categories may be especially vulnerable to certain risks, such as living near environmentally hazardous facilities and being exposed to air pollution (Bullard 2000). Analyzing and documenting such variations in risk exposure is an important task for environmental sociologists.

Scholars in the field of intersectionality studies are usually careful to avoid a deterministic perspective (Olofsson et al. 2016). Indeed, people are not marionettes, but have agency and reflective capacity. We cannot deduce exactly what and how people will think, feel, and act just because they occupy a certain position in society. People are neither fully free nor fully determined, but are somewhere in between: strongly influenced by intersecting social determinants, but also with the capacity to reflect and act based on their own choices.

An intersectional perspective is useful for environmental sociology, because it provides lenses through which we can understand how certain groups are marginalized and experience marginalization. Sometimes – as is often the case with indigenous peoples – groups are rendered completely invisible in policies, campaigns, and discourses (Olofsson et al. 2016). They do not even exist in the minds of those who make decisions that fundamentally shape their lives, such as opening new mines, building oil and gas pipelines, and burning rainforest to create farmland for cattle. These things often happen in regions populated by indigenous peoples or others who live far from the centers of power and have little ability to influence decisions, even when they concern their lives and livelihoods. It is therefore important to ask questions such as: which actors or groups of people are represented in narratives of environmental problems and solutions? Who are considered to be legitimate change-makers and leaders? Who is privileged and who is subordinated or even made invisible? What opportunities do they have to speak and act on their own behalf? To answer such questions, we also need to include perspectives on power (see Chapter 2) and participation. We will return to these and related questions later. First, however, we need to look at another important environmental area in which the distribution of environmental problems is central, namely environmental justice.

Environmental (in)justice

Environmental justice is one of the main perspectives used in studying social stratification within the environmental social sciences. The theory of environmental justice addresses the exposure of people and groups to environmental pollution and risks. Campaigns against environmental injustice have taken place around the world. A classic study is *Dumping in Dixie: Race, Class and Environmental Quality* by Robert Bullard (2000). In this book, Bullard shows how, in the 1960s, the geographic connections between sites of noxious facilities (e.g., waste disposal sites) and highly segregated areas in the southern USA sparked a protest movement against environmental racism that brought together the civil rights and environmental movements.

The field of environmental justice focuses on how environmental "goods" (such as access to clean air, clean water, food, natural resources, green urban areas, nature preserves, and aesthetic surroundings) and "bads" (exposure to pollution, environmentally detrimental facilities, toxic waste, noise, substandard housing, and risk of flooding) are unequally distributed along the lines of class, racial/ethnic groups, gender, income, and other categories (Agyeman et al. 2003, DesJardins 2006, King 2020, Pellow 2020, Roberts et al. 2018). These studies often have a geographical dimension. They often focus on a particular territory and a socially segregated local community, and examine how the socio-material environment (facet 3), industrial exploitation, and pollution affect that territory and community. Such studies reveal how segregated, poor, racialized, and marginalized communities are disproportionately affected by environmental contamination. Other studies focus on the global scale, addressing for example global climate justice and how some places are able to develop protective strategies, while other, often more vulnerable and densely populated places lack the resources to plan for future climate change. Another topic is the study of disasters, which looks at how the effects of natural disasters such as floods and hurricanes affect different segments of society in different ways.

Environmental justice can be seen as the study of the interactions between a differentiated society and a differentiated environment (Roberts et al. 2018). There is no such thing as a healthy environment for everyone, and a poor, polluted environment is mainly left to marginalized groups with few opportunities to choose where to live. Access to clean air, clean water, clean soil, open spaces, urban parks, nature preserves, and so on, is extremely unevenly distributed, with some social segments having abundant access to them and others having hardly any access at all. Such processes of exclusion are sometimes the visible result of specific decisions, but more often they are the outcome of more mundane processes linked to general patterns of social stratification. The most important mechanism of exclusion is economic (intersecting with categories such as ethnicity, immigration, racialization, etc.). Those who can afford to live in the more attractive and environmentally healthy areas do so. Wealthier segments of the population are more mobile and can move elsewhere if they wish; the phenomenon of "white flight" is well documented in research (Kruse 2005). These segments also have the resources (including knowledge) needed to adapt to changing environments, for example by installing air purification devices.

Also, things that are considered to be truly global phenomena, such as climate change and biodiversity loss, have very diverse and uneven consequences. The recent experience of the Covid-19 pandemic clearly shows that global phenomena hit different segments of society in very different ways (see Box 3.1).

Box 3.1 Covid-19 and environmental justice

Around the world, the Covid-19 pandemic prompted social interventions and mobilizations of state power unheard of outside of wartime (Lidskog et al. 2020). Entire cities were locked down, international borders were closed, and the web of human activity (outside of digital networks) nearly ground to a halt. In early April 2020, one-third of the world's population was under some form of movement restriction or lockdown (Koh 2020). Today (2023), it is estimated that there have been approximately 700 million cases of Covid-19, resulting in more than seven million deaths. Nearly 13 billion doses of vaccine have been administered.

But the deaths and ill-health caused by the pandemic were unevenly distributed, with factors such as age, class (not least in terms of housing standards), and working conditions making some more exposed and vulnerable to the virus (Brown and Zinn 2022). Not only did the pandemic itself kill millions of people in a very uneven way, but so did government responses to the pandemic.

Government measures to limit the spread of the virus have not only saved lives, but have also had serious negative and unintended consequences, causing or exacerbating broader social problems such as lost schooling due to school closures, increased unemployment, social isolation, and difficult economic and social conditions for populations. These effects are extremely unevenly distributed, both geographically (across different countries) and socially (across different social categories).

How environmental injustices are manifested in the operations of extractive industries is a topic that has received increasing attention in environmental justice research and environmental sociology (Malin et al. 2019). The extraction of oil, coal, gas, metals, minerals, and precious stones raises many issues of justice and rights. Systematic patterns of land grabbing, inequality, and unemployment are often associated with extraction sites, especially when they are located in developing countries. Local communities that are economically dependent on extractive industries, which are often run by powerful multinational corporations, often have very little capacity to mobilize public resistance. Indeed, even in wealthier countries, persistent poverty is a consistent structural problem in communities that host extractive activities (Malin et al. 2019).

This also points to the dynamic and complex character of many social processes. Current efforts to achieve sustainable transportation have led to electrification in many countries, resulting in an increased need for precious minerals (lithium and cobalt) to produce vehicle batteries. This has created a new demand for mining, which often takes place in locations with poor working conditions and also causes local environmental problems. In practice, this means that while some countries are improving their environmental situation by electrifying vehicles (and thus avoiding air pollution from transportation), other countries are facing a worsening environmental situation due to mining activities (Henderson 2020). In this sense, one country's "greening" is another country's pollution, but this is often covered up and is rarely touched upon in policy debates. Thus, unless issues of environmental justice are considered, "clean transportation" may be clean for some social segments or geographic areas, but anything but clean for the other segments and places that make this greening possible.

Why does environmental injustice still exist?

Research proposes various and interrelated explanations for why environmental injustices are reproduced. Roberts et al. (2018) distinguish between three different kinds of explanations – economic, socio-political, and racial discrimination – which mainly focus on institutional factors (facet 5). Economic explanations focus on market dynamics and the logic of capitalism. The spatial concentration of polluting industries is a result of capitalist market forces that force firms to cut costs and seek cheap land and labor. This is also the main reason for the outsourcing of production to developing countries, which leads to complex, global commodity chains (Brand and Wissen 2021, Fridell 2019). Localized industrial exploitation and pollution have the unfortunate consequence of driving down property values in an area. This means that the people who work in the area can afford to live there but nowhere else, and no one else would find it attractive to live there. Socio-political explanations complement this perspective by focusing on the role of political power. In areas with poorer and less educated populations, it is often easier to gain political support for local exploitation. Public resistance is less likely (see the three dimensions of power, Chapter 2). Explanations based on racial discrimination add the dimension of institutional or structural racism. Institutional racism does not necessarily take the form of overtly racist attitudes and viewpoints being expressed in public planning, "because racial privilege and disadvantage are built into the society's social structures and systems of governance and commerce" (Roberts et al. 2018, p. 242). In practice, these three dimensions are inextricably linked and mutually reinforcing.

To these explanations we need to add a *relational* perspective. In their article "What Is Environmental Racism For?" Seamster and Purifoy (2021) argue that research on environmental justice should place more emphasis on how relationships between different communities contribute to reproducing injustices. The development of one place – such as a "white town" – depends on exploiting the resources of another place – a "black town" – which results in infrastructural inequality. The richer community can flourish and become overdeveloped at the expense of neighboring communities that are also economically and symbolically devalued by racializing processes. Lower income groups have no choice but to live there, and the wealthier classes can keep their town for themselves. The appropriation of resources can be achieved and sustained through rather mundane processes of economic transactions and political exclusion, that is, the denial of democratic participation or the erosion of the agency and effectiveness of political representation. The example above about how the electrification of transportation has led to the exploitation of mining communities is a case in point. A country deliberatively and democratically decides to upgrade its transportation infrastructure in a way that causes environmental destruction to other countries, places, and people.

Thus, we ought to combine economic, political, cultural, and socio-psychological aspects in the analysis, and keep the focus on the relational dimension. While the broad explanations that focus on economic and political power primarily address the fifth facet of our model of the social (facet 5), it is important to see how these mechanisms also operate in and through socio-material arrangements (facet 3), and through people's social relations (facet 2) and inner lives (facet 1), as people adapt socially, cognitively, and emotionally to the circumstances of their local environments. The intersectionality perspective reveals that several aspects of domination interact to produce environmental injustices. Malin and Ryder (2018) note that environmental injustice often intersects with "dominant ideologies that operate as classist, racist, sexist, nativist, ableist, homophobic and anthropocentric matrices of domination" (p. 1). Multiple forms of oppression and inequality interact to shape and reinforce environmental injustice.

Another key topic in environmental justice research is the highlighting of different aspects of justice (see Kopnina 2014, Malin et al. 2019, Roberts et al. 2018). We can distinguish between forms of justice in terms of their distribution within stratified society (between social classes/groups/disadvantaged populations) and even extend this to include the larger *biosphere* (between human and non-human species). We can distinguish between *spatial* (intragenerational such as between countries/regions) and *temporal* (between present and future generations) justice. In addition, we can focus on the content of justice in terms of the distribution of environmental "bads" and "goods," or we can address procedural, corrective, and more structural issues. A more detailed account is given in Box 3.2.

Box 3.2 Forms of environmental justice

As a field of study, environmental justice is concerned with the social distribution of environmental benefits and burdens. Research has shown that society often places disproportionately high burdens on those who are most disadvantaged, while the benefits accumulate elsewhere.

1. **Distributive justice** refers to the perceived fairness of outcomes or resource allocations (in terms of both benefits and burdens). Environmental justice is concerned with the fair distribution of environmental "goods" and "bads" (see discussion above). We can distinguish between three subcategories.

   a. **Intergenerational justice** refers to the right of future generations to have their needs and desires satisfied in the same way as present generations. Its best-known formulation is in the Brundtland Report: "sustainable development is development that meets the needs of the present without compromising the ability of future generations to meet their own needs" (UN 1987, p. 41). This is the temporal dimension of justice.
   b. **Intragenerational justice** refers to present generations, and is concerned with such things as the distribution of environmental "goods" and "bads" among countries and regions, as well as among social segments within a society. This is the spatial dimension of justice.
   c. **Animal justice** refers to the inclusion of non-human beings in the moral community. Non-human species should be included in questions of justice because they are living and sentient beings that have the right to develop and flourish (a position that is sometimes referred to as biospheric egalitarianism).

2. **Procedural justice** concerns the perceived fairness of rules and processes used to determine outcomes. In environmental justice, this often includes the ability of people to meaningfully influence decisions that significantly affect them. For example, workers and residents affected by industries should have democratic means of influencing their operations, such as when corporations investigate sites for gas or mineral extraction. Residents should be given the power to influence whether or not these industries should be there, issues of safety and protection, etc. This also involves "recognition," that is, identifying those social groups and worldviews that have a legitimate right to be represented, and thereby counteracting the (intentional or unintentional) silencing of groups.
3. **Corrective justice** refers to adequate remedies and compensation for communities that have been harmed by environmental contamination.

Remedies can take the form of economic compensation, but can also include the possibility to reclaim and restore a damaged landscape. Unfortunately, restoring a place or community is often not possible because of irreversible damage, such as loss of life or chronic illness, the extinction of a species, or the elimination of a local ecosystem.

4. **Restorative justice** is closely related to corrective justice, but refers to holding offenders accountable for understanding the harm they have caused, giving them opportunities to redeem themselves, and discouraging them (and others) from causing further harm. Restorative justice also gives victims the opportunity and power to speak about what happened to them and how it has affected them, and requires the offenders to listen.
5. **Structural justice** refers to broader issues related to fair opportunities for work, income, education, living conditions, etc., and thereby counteracts systematic disadvantage that may affect social segments of people. Structural justice is a prerequisite for other forms of justice.

Expanding environmental justice: The global and temporal scales

The literature on environmental justice has progressed in recent years. While studies of environmental justice have often focused on the local or regional scale (e.g., local sites of environmental issues), it is equally applicable to other scales, for example, considering post-colonial global relations as well as justice along global commodity chains. Climate justice is one such key topic. Climate justice scholarship draws attention to global inequalities and observes that societies are divided into those that cause the problem of climate change, such as wealthier nations, the fossil fuel industry, and the rich, and those that suffer the most from its effects: poor people, vulnerable places, and developing nations. Smaller and poorer nations are often disproportionately vulnerable to sea level rise, droughts, floods, and hurricanes (Roberts et al. 2018, p. 236). Clearly, in our globalized postcolonial world, what happens between nations is critically important. We will briefly mention a few key topics and schools of thought.

One is the tradition of theories, often of a Marxist kind, that link an analysis of capitalism to the ways in which historical imperialism and colonialism still structure global relations. We introduced some of these theories in Chapter 2 and continue the discussion here. Global capitalism has created an international system of core and peripheral countries, in which the latter are an important source of natural resources for the former. This sociological tradition culminated in Immanuel Wallerstein's famous world-systems theory (Wallerstein 2004). Global capitalist society is not just a class society, but is a society in which capital extracts resources and profits from developing countries without giving them an equivalent value of resources in return. More resources flow out of these countries than flow in (Armstrong 2019). Some theories use the concept of "metabolism" to stress the unsustainable exchange (metabolism) between society and nature in contemporary society (see Chapter 2). These studies show how extraction in one place and wealth accumulation in another amount to an "unequal ecological exchange" (see review by Davidson 2018). When a natural resource is sold from a rural and developing country to a more developed one, there is a large net gain for the latter. Underdeveloped countries are systematically robbed of their wealth. Stoddard et al. (2021) show how "global extractivism" is closely connected to geopolitics and militarism. Ecologically unequal exchange is predicated on historical and contemporary colonial forms of extractivism. Moreover, militarism has long played a direct but relatively neglected role in causing climate change, and also leads to mindsets that "facilitate wider ideologies of control, which form an important part of the inertia against zero-carbon transformations" (p. 663).

Another type of literature, which sometimes overlaps with the above theories, focuses on global value/supply chains. These studies show how economically powerful actors – often brand-owning multinational corporations such as Nike, H&M, IKEA, Walmart, and Monsanto – are able to control and accumulate economic value from suppliers in developing countries (Boström et al. 2015, Fridell 2019). The globalization of commodity chains has been driven by cost-reduction incentives. Considerable outsourcing of domestic production to developing countries with lower wages and lax social and environmental regulations has been the pattern in many sectors, such as food, textiles, and electronics. Many consumer goods literally travel around the world, making it extremely difficult for any particular actor to be informed about what is happening behind the scenes and to assess the economic, environmental, and social sustainability impacts. Indeed, many of the commodities that people in rich nations enjoy are imported, which leads Brand and Wissen (2021) to speak of an "imperial mode of living" in the global North (see Chapter 2). The average consumer is not even remotely aware of the social and ecological costs required to produce and distribute these items, and that these costs are borne by people and communities in far distant places.

The sustainability discourse highlights the moral imperative that future generations should have the same chances to realize their wants and needs as the now-living generations. Until recently, studies of environmental justice have primarily focused on equity between currently living generations, but the consideration of climate change explicitly or implicitly gives more attention to intergenerational justice. Greta Thunberg, the initiator of the international movement *Fridays for Future*, has consistently stressed that the older generation has caused the current climate crisis and is now passing it on to her generation to solve. As she puts it: "We talk about our future, they talk about their present." Thus, apart from emphasizing geographic inequalities (the spatial dimension), the environmental justice literature has increasingly addressed the temporal dimension (the future of younger and future generations).

The temporal perspective concerns not only the future, but also the past. Elizabeth Hoover (2018) has introduced the term "reproductive" to environmental justice, as in the concept of *environmental reproductive justice*. This considers a community's historical possibilities to sustain and reproduce itself. We ought to ensure that entire communities, including their particular forms of social life and cultural patterns, are not eroded, but are given opportunities to reproduce and develop. The addition of this reproductive aspect to the concept of environmental justice is very important, because what is often at stake is the very survival of groups of people, for example indigenous communities. In her article, Hoover focuses on one Native American community, the Akwesasne:

> Sites ranging from industry to mines to military bases, as well as places impacted by the release of pesticides and other agricultural by-products, negatively affect not only the surrounding environment, but also the health, culture, and reproductive capabilities of the Indigenous communities they border. (Hoover 2018, p. 9)

The concept of reproduction (or reproductive) is apt, because it is not limited to the present situation. It includes both history as well as the future. To be able to reproduce is to have a cultural relationship with the past and to recognize the right to maintain existing cultural patterns. It also implies traditions and local knowledge about how to sustain that environment. This question of sustaining ways of life, including the social life of certain groups in society, has affinities with another concept, ecological violence.

Ecological violence

Some scholars argue that environmental injustice is essentially a matter of violence, because some of its consequences are so severe. As Stewart Lockie puts it: "environmental injustices are injustices not because they reflect inequality alone but because they shorten peoples' lives, compromise their physical and mental health, and increase their vulnerability to other sources of risk" (Lockie 2018, p. 177). Thus, violence is even part of the definition of environmental injustice: "I define environmental injustice as violence perpetuated on the bodies, minds and livelihoods of its victims through the chemical and biological pathways of ecosystem processes and human metabolism" (Lockie 2018, p. 178).

Environmental disasters occur around the world and are frequently reported by the news media. These types of phenomena lend themselves very well to the format of news media, which publish dramatic images of how unstoppable storms and fires destroy cities (Berglez and Lidskog 2019). The media publish stories about how local communities are severely damaged and even cease to exist. Disasters are often portrayed as coming out of the blue; a local community, city, or region is suddenly hit by a storm, flood, or drought. In his book *Slow Violence and the Environmentalism of the Poor* (2011), Rob Nixon shows that this is a far too limited understanding of environmental disasters. Many disasters are in fact "slow"; they occur gradually, their effects are distributed across time and space, and they are conditioned by powerful and deep forces that are often neglected in news coverage. The reason for this neglect is that these kinds of disasters do not fit the media logic; they unfold slowly, with no immediately visible effects, and it is often difficult to find one single actor that is responsible. Microplastic waste in the oceans, ocean acidification, and air pollution are all examples of environmental disasters whose major effects appear gradually and are rarely framed as disasters.[2]

We therefore need to expand our conception of environmental disasters to also include slow violence, those environmental disasters that are not spectacular events localized in time and space, but instead cause a gradual and creeping accumulation of damage and do not fit the media's dramatic packaging equally well. Nixon also shows that this kind of ecological violence is being outsourced to the global South and to future generations, whose environments are gradually being toxified. This is slowly eroding the prerequisites for life of people and communities. Environmental disasters have extremely uneven impacts. Disempowered people, communities, and countries are the hardest hit and lack the resources and capacities to cope.

These environmental disasters can be observed and responded to, and numerous stories are told of local communities that are suffering from gradual environmental destruction (Davies 2022, Erikson 1995). The problem is that these stories are not treated as if they really matter, and thus slow violence is rendered invisible to broader segments of the public. It is therefore extremely important to reveal and emphasize the connections between cause and effect, now and future, and here and there, in order to legitimize and disseminate these stories of how local communities are being hit by ecological violence. In doing so, it is important to have a transnational perspective on disasters, because their causes are external to the affected communities and places, and at the same time there is a great need to listen to those who live in the midst of these environmental disasters and experience daily toxic exposure.

Above, we highlighted the concept of *environmental reproductive justice*. This too can be linked to violence. Bacon (2019) proposed using the concept of *ecological violence* to address colonial aspects related to exploitation of the social lives of indigenous peoples. If groups of people or communities are unable to reproduce themselves because of another group's domination, it is appropriate to call this power a form of *violence*. Several historical elimination practices may be involved with such ecological violence:

- Physical elimination, through genocide.
- Land appropriation, redistribution, privatization, pollution, erosion, and renaming.
- Cultural elimination, such as suppression of indigenous cultural awareness in schools.
- Discursive under- and misrepresentation, such as renaming/repurposing culturally significant places.
- Neglect of political rights and democratic representation.
- Criminalization of traditional practices of hunting, fishing, burning, etc.

Bacon is referring to settler colonialism in North America, but elimination practices appear throughout history and around the globe. The Amazon rainforest and the islands of South Asia are important examples of places where the traditional livelihoods of indigenous peoples are seriously threatened.

As highlighted in Box 3.2, a particular form of environmental justice concerns non-human species. This is represented not least by the animal rights movement (or animal liberation movement), which is critical of the strict demarcation between human and non-human animals, and to the attribution of only instrumental value to animals. The movement struggles to end violence against animals, for example in the food and clothing industries and in research.

Social stratification and disasters

It is important to consider environmental justice when seeking to understand and analyze environmental disasters. Environmental disasters are nothing new in history, but climate change, land use, urban planning, population patterns, and other factors are making environmental disasters more frequent and severe today (Lidskog and Sjödin 2016). It is important to note that regardless of whether a disaster occurs slowly or suddenly, it has multiple consequences that unevenly impact people and places. Because society is socially stratified, it is also geographically stratified, which means that people living in different places are exposed to disasters differently and also have different resources to cope with them. For example, sections of cities inhabited by the urban poor have become the areas most vulnerable to flooding (Marks 2018).[3]

When Hurricane Katrina hit New Orleans in late August 2005, it caused 1,800 deaths and enormous economic costs. What may be less known is that no environmental disaster has led to a larger internal US diaspora of displaced people – 1.7 million people (Brunsma et al. 2007, p. xv). More than 400,000 people became homeless. In the book *The Sociology of Katrina* (Brunsma et al. 2007), environmental sociologists explore this disaster and show how it affected different segments of the population unevenly. The book also shows that the severe consequences of the hurricane could be amplified or weakened by existing social structures in the community. For example, many of the victims were already dependent on social support, and many of the damaged houses were already dilapidated before the hurricane struck and flooded them. This study shows not only that some groups were more severely affected by the disaster than others, but also that inequalities were exacerbated by the hurricane and its aftermath. The city was rebuilt and urban areas were redeveloped, but many people did not move back. It is even possible to say that the hurricane is not over, because the damage is still there in terms of evacuees who never returned. As the book states, many residents witnessed three major failures of social institutions: a failure to prepare, a failure to respond, and a failure to rebuild (Brunsma et al. 2007, p. 12; see also Malin and Ryder 2018).

Segregated effects are not unique to the Katrina disaster. For example, when the storm Sandy hit New York in October 2012, it caused flooding of the lower levels of houses and apartment buildings, leaving 800,000 residents without power, heat, or hot water for up to several months during the winter season (Medwinter 2020). Medwinter's study focused on the Rockaway peninsula, which has a rich (mainly white) part and a poor (mainly non-white) part. Those living in the rich part were able to utilize their stronger social capital and networking skills to more easily access informational, material, and emotional resources from both institutional (governmental and non-governmental organizations) and more informal sources. The study shows how the different facets of the social interact. Accordingly, racialized processes, poverty, and class inequality can significantly shape not only what happens during a disaster, but indirectly also the aftermath during the long process of relief and recovery. Recovery has both a socio-material and an emotional (inner life) dimension. After a disaster, it is obvious that some basic material necessities must be provided to people – food, water, medicine, shelter, clothing, protection from the cold, and so on. But we cannot underestimate the importance of other aspects of the social – the ability to forge "affective bonds." Applied to our model of the social, facet two (social relationships) and facet one (inner life) are both critical. People's ability to forge affective bonds with others – friends, neighbors, volunteers, firefighters, government aid workers – may be critical to their ability to cope with stress, anxiety, frustration, grief, and loneliness. Perceiving that help is within reach is vital.

The above studies stress the importance of paying attention to distributional issues before, during, and after disasters. It is vital not to lose sight of the causes of the disaster. As one scholar of disaster studies put it: "for six decades, disaster research has been preoccupied with the study of what happens during and after disasters, rather than asking why disasters happen" (Tierney 2012, p. 66). It is important, as the environmental justice literature stresses, not to see disasters as caused by "blind forces." Rather, they are often interconnected with other factors such as poor planning, shoddy technology, low crisis preparedness, and economic incentives.

Representation and democratic participation

A key aspect of environmental justice concerns procedural issues: democracy. Do those affected by an environmental problem have access to the resources and opportunities they need to be able to represent themselves and communicate their concerns? This is a classic problem of participation and representation, and it is particularly challenging in the case of the environment, because those most affected often poorly match the population that has formal and practical opportunities to take part in democratic elections and decision-making. This is the case for many of the most serious environmental problems, which have global, long-term, and even irreversible consequences. As stressed above in Box 3.2, environmental problems are transboundary in nature. They involve justice between different regions of the world, justice between generations, and justice between species (Lidskog and Elander 2007). This means that they challenge much of traditional democratic decision-making (for various forms and challenges of "environmental democracy," see Fischer 2018).

Unlike with many social issues, the environment can never represent itself. A blind person may represent herself as a blind person, or represent other blind people when speaking about the difficulties that blind people face in society. But an environmental problem must be represented by actors (representatives) that act and speak on behalf of the environment (Boström and Uggla 2016). This ambiguity of representation is particularly salient when the issue concerns constituencies that cannot speak for themselves. In conventional "representative democracy," representation is limited to national citizens. However, many environmental problems are transboundary in nature; they cross spatial and temporal boundaries.

Those responsible for causing pollution and other environmental problems may live and operate abroad, outside the boundaries of national democratic institutions. Huge numbers of people lack citizenship status in the country where they live, and climate change is expected to cause migration flows that will result in growing numbers of non-citizen residents. Another key missing constituency is future generations, who can only be represented indirectly, for example by expert organizations and scientific claims (such as the IPCC or Future Earth) or by actors speaking on their behalf (such as *Fridays for Future*). Missing constituencies that cannot speak for themselves also include a wide range of environmentally threatened entities, such as non-human animals (e.g., endangered species), habitats, ecosystems, or planet Earth as a whole. Various procedures, tools, institutions, and artifacts need to be developed to represent relevant non-speaking entities in decision-making processes. Thinking in terms of representation is never a straightforward matter, but at the same time it is always necessary. Anyone can claim to represent the environment in one way or another, and invoke certain experiences, experts, scientific findings, identities, forms of ownership, authorities, or other principles to support and justify one's representational claim (Boström and Uggla 2016, Boström et al. 2018). Given this inevitable ambiguity, it is important to openly discuss who is considered a legitimate expert, representative, participant, "stakeholder," affected member of the public, and so on, as this is itself a matter of social stratification and involves power. For example, historically, men have claimed the right to speak on behalf of women, as is still the case in many countries.

Even when affected publics (constituencies) have opportunities to communicate and represent their opinions, power imbalances and empowerment remain very important issues. Participation in politics and public policy is always a challenge in terms of equality, balance, and effectiveness. Initiating an environmental assessment or decision-making process to involve more groups from the public, such as residents of a community where a corporation is planning industrial activities, does not automatically make the process more "just" or "fair," because the already powerful actors may be the ones who set the agenda, compile evidence, and are able to advance effective arguments (Busca and Lewis 2015). An inclusive process may even further strengthen powerful actors, including local elites. The question is whether weaker actors can be empowered in various ways and exert influence. Issues of participation and representation cannot therefore be meaningfully discussed unless they are guided by a proper understanding of power relations and power dynamics (see Chapter 2 on the different dimensions of power). We will return to this topic of mobilizing counterpower in Chapter 6.

Next, we will consider another aspect of social stratification and the social distribution of environmental problems: income inequality.

How inequality causes environmental problems and prevents action

Chapter 2 discussed the causes of environmental destruction. In this chapter we need to add the fact that inequalities themselves are a factor in causing environmental problems. It is therefore necessary to discuss how both wealth and poverty can contribute to ecological destruction, and how high levels of inequality prevent collective action and the taking of responsibility.

Wealth and poverty

We live in a highly unequal world. At the time of writing (the figures are rapidly becoming outdated) the richest 10% have been responsible for about half of all global carbon dioxide emissions since 1990. At the same time, the poorest 50% account for less than 10% of emissions. The richest 1% emits twice as much as the poorest half (for the figures, see Chancel and Piketty 2015, Kartha et al. 2020, Kenner 2019, Oxfam 2020, Stoddard et al. 2021).

A key determinant of life chances is country of birth. In his book *Why Global Justice Matters*, Chris Armstrong goes so far as to claim that, "It is a far bigger determinant of our life chances than our class, race or gender. Inequalities *between* countries far outstrip inequalities *within* countries" (Armstrong 2019, p. 13). He found that the average US citizen is responsible for about 165 times as much emission of carbon dioxide per year than the average citizen of Chad, Rwanda, or Ethiopia (ibid, pp. 13–14). Thus, it would take 165 people in those countries to equal the carbon emissions of one US citizen.

People born in the developed world tend to take their privileges for granted. They inherit and internalize a set of advantages and associated value systems that seem normal to them. Wealth implies the reproduction of unsustainable, high climate impact lifestyles and ecological footprints (see Akenji et al. 2021). People socialized in affluent societies are firmly embedded in ways of life that are deeply unsustainable and incredibly difficult to break out of (Boström 2023). All five facets of the social are at work here, embedding and entangling people in circumstances that are hard to influence, at least for individuals. We have been socialized to have certain expectations about what a good life is and what kinds of things are worth striving for. Our group membership and social status push us to engage in certain activities and to acquire and use certain possessions, and society is organized in ways that limit the possibility of living more ecologically sustainable lives.

As a group, the superrich have received increased attention in both public debates and scientific work, due to their extreme levels of environmental emissions. Dario Kenner (2019) uses the term "polluter elite" to designate the richest 1% of the population in countries such as the USA and UK, who tend to have luxurious lifestyles with extremely high carbon footprints. They engage in status consumption by owning mansions and multiple properties, and traveling frequently to exclusive destinations by yacht, private jet, sports car, etc. They also have the resources to adapt to and escape the adverse effects of climate change and ecological devastation. They own residences in different regions, and can move between them to avoid droughts and heatwaves. They are very powerful, have investments in the fossil fuel industry, and can shape political agendas by providing resources to lobbyists and conservative think tanks. Some of them even control political agendas by occupying high political positions in societies (even as presidents and prime ministers).

An additional problem associated with the lifestyles of the superrich is that their actions and messages spread and defend consumerist norms, which other less wealthy populations pick up as inspirational ideals (Otto et al. 2020, Kenner 2019, Wiedmann et al. 2020, Boström 2023).

One example of this is air travel, an activity that is extremely unevenly distributed between and within countries. Emissions from international aviation more than doubled between 1990 and 2017 in countries such as the USA, UK, Sweden, Italy, Germany, and the Czech Republic, more than tripled in countries such as Spain and Australia, and increased even more in Luxembourg, Turkey, and Iceland (Gössling and Humpe 2020). While a large share of the population in these wealthy countries fly infrequently, perhaps once a year, it is still the case that only 11% of the world's population traveled by air in 2018. A small minority of people in these countries are high emitters through frequent flying. 1% of the world's population is responsible for more than half of global CO_2 emissions from air travel. Travel and hypermobility are linked to status and play an important role in social media communication. The explicit and implicit messages of wealthy celebrities (influencers, fashion designers, business leaders, musicians, sports stars, etc.) are often linked to travel and other activities with high climate and ecological impacts, and they reach millions of followers. "Celebrities define aspirational lifestyles and desirable consumption" (Gössling 2019, p. 2).

The flip side of wealth is poverty. Severe and growing inequalities characterize the world. According to the World Bank (2023), around 700 million people live in extreme poverty with poor access to food, education, and healthcare. Although the situation has improved since the 1990s (Armstrong 2019), the numbers are still extremely high. The problem is challenging to address and is exacerbated by pandemics and wars. According to the World Food Program (2023), more than 900 million people are struggling to survive catastrophic hunger (one step away from famine) in 2023, ten times more people than five years earlier. Poverty is not just a lack of income; it is a lack of many of the elements of a secure and decent life, and it is usually accompanied by a lack of adequate democratic and political representation. Clearly, poverty is a killer, both directly and indirectly (Armstrong 2019).

Poverty also means an inability to deal with environmental problems. The need to ensure one's own survival takes precedence over everything else. It may be necessary to be employed by multinational corporations and to work at sites where nature is exploited and natural resources are extracted. Poverty makes it impossible to plan, invest, and act in the long term.

It is important to keep in mind that wealth and poverty are structurally related, as we argued above when discussing environmental justice. Within just a few generations, many welfare societies have been able to leave poverty behind. This is partly due to historical patterns of colonialism and imperialism and more contemporary patterns of global capitalism, which mean that resources flow from developing to developed countries much more than the other way around (through development aid, etc.) (Armstrong 2019, Brand and Wissen 2021). Thus, the world faces not only the problem of enduring poverty, particularly in Africa, but also the problem of growing inequality and power asymmetries. Poverty and inequality should therefore be considered common global problems, rather than local ones. One reason for this is moral: we are all human beings. No one can be held responsible for being born in the wrong country (or for their gender, sexuality, class, and skin color). Another reason has to do with collective problem solving (Stoddard et al. 2021). Inequality results in collective dilemmas. Armstrong (2019, p. 22) exemplifies this with climate change:

> We have a classic stand-off. Rich countries quite rightly point out that the world cannot afford for poor countries to industrialize along the same dirty, carbon-intensive path they pursued themselves. Poor countries quite rightly point out that, pressing problem though climate change is, their own endemic poverty demands their attention. And as it stands, industrializing as fast as they can – even if this means dirty and dangerous industrialization – appears to be the best way to make progress.

Inequality has negative effects in that our institutions – global, national, and local – tend to be much more responsive to the preferences of the rich, and much less so to those of the poor. Institutions such as the WTO, the IMF, and the World Bank are designed primarily by the elites of richer nations. Another very harmful effect, as suggested in the quote above, is that massive inequality makes it more difficult to solve major global problems. Developing countries face their own pressing and immediate problems, while those who can afford to pay for the solutions make excuses and find ways not to do so. A more equal world would mean less conflict and greater potential for truly democratic decision-making, and more effective handling of common problems. We will return to the topic of international collaboration in Chapter 6. Now, in the final section of this chapter, we will take a closer look at an important sociological concept: relative poverty.

Relative poverty

In many cases, the problem is not one of absolute wealth or poverty. Often, it is the gap between rich and poor countries, or between rich and poor people within a country, region, or local community, that has serious negative environmental and social impacts. Inequality is at the root of a wide range of social problems (Wilkinson and Picket 2011).

Inequality – which is commonly associated with income and economic wealth – has increased overall in welfare societies in recent decades (Isaksen and Roper 2008, Jorgenson et al. 2019). One reason is that the rich are getting richer, while the poor are staying at the same income level or becoming even poorer (Sayer 2015). In addition to returns on savings and investments, those who already have resources benefit from all sorts of subsidy policies. On the time scale of human history and prehistory, our current societies appear highly unequal and exceptional in terms of wealth. Unfortunately, the growing inequalities in terms of income and economic wealth give rise to many serious social and psychological problems.

As the American sociologist Robert Merton (1938) noted more than 80 years ago, behaviors labeled "deviant" (such as crime) are produced by a combination of limited resources (poverty), limited opportunities, and a shared system of success symbols within a society. Merton considered such deviant behavior to be normal in a cultural context such as American society, where pecuniary strength is considered highly prestigious, but the legitimate institutionalized means to obtain these ends are highly unequal. His discussion gave rise to the idea of *relative poverty* or *relative deprivation.*

Accordingly, in sociological studies of poverty it is often more relevant to speak of relative poverty than of absolute poverty. What matters most, at least in welfare societies, is that members of different groups compare themselves with other groups and feel injustice and envy. This has been the case so far because there are common goals, as Merton observed. A more vertically differentiated social structure can lead to a larger proportion of the population feeling worse off and dissatisfied. If you have little, it feels more bearable if you share this situation with everyone else (or most people) in your society.

In terms of wealth and income, the social structure of Western societies has become more unequal in recent decades. This is discussed by the social epidemiologists Richard Wilkinson and Kate Pickett in two books (2011, 2018), in which they convincingly show how important relative poverty is in relation to many health and social problems. They relate comparable statistics on the income gaps in 23 rich and democratic countries to various measures of social and health problems, and show how important the size of the income gap is (and that absolute levels of income are unrelated to the same problems). They found correlations with the quality of social relations in a society (trust), physical health and life expectancy, obesity, mental health (increases in anxiety, depression) and drug use, educational performance, teenage parenthood, violence, and incarceration, for example. They also found that higher inequality reduces social mobility and makes the social structure more rigid. Hence, it becomes more difficult to move up and down the social ladder when inequalities increase.

Wilkinson and Pickett (2011, 2018) use a wide range of sociological explanations: social status, comparisons, social judgment and evaluation, peer pressure, and so on. It is a question of relative poverty, as Merton and other sociologists have discussed. What matters is where we stand in relation to others in the same society. Inequality gets under our skin, the authors stress. It deeply affects our inner life (facet one). Indeed, human beings are extremely sensitive to inequality and injustice. Wilkinson and Pickett argue that social anxiety and insecurity are the most common sources of stress in modern societies, and that these are deeply related to social comparison, status judgment, and peer evaluation. Differences in status and wealth create a social gulf between people. This social gulf is a root cause of excessive individualism, and it destroys trusting relationships, undermines a culture of cooperation, and spurs dominance strategies such as self-advancement and status competition (facet two). It is not only those at the bottom who are harmed by inequality. The vast majority are harmed by greater inequality, because the gap itself increases anxiety, insecurity, and a competitive culture of comparison, status-seeking, and mutual judgment. Everyone fears falling down the social ladder.

At first glance, these issues of social justice seem to have nothing to do with the environment. However, the authors emphasize that inequality is not only a root cause of social problems (such as public health), but also has serious environmental consequences. First, it spurs overly individualized, status-oriented, and excessively consumerist lifestyles that reproduce high ecological and climate footprints. This aspect of how excessive consumption is linked with social status and comparison is well documented in the sociology of consumption (see Boström 2020, 2023).

Secondly, it destroys trusting relationships, collaboration, and a sense of shared responsibility. More equal societies appear to foster social trust and a greater sense of collective responsibility, which in turn are crucial for public legitimacy and the ability to mobilize joint political action to address climate change (Stoddard et al. 2021). Participation in community life requires a certain level of trust. Broken trust and sense of collective responsibility lead to collective dilemmas. Shared responsibility becomes difficult: "You have more money and power, so you are responsible for..." or "Why should I reduce my consumption, which has such a small environmental impact compared to rich people's excessive consumption?"

Thirdly, the issue of higher social inequalities may also divert attention from environmental problems and keep them off the agenda. Inequality generates many social problems, which may support the view that we need to address social problems before we can deal with environmental problems, when in fact social and environmental problems are often interlinked, as the environmental justice scholarship shows.

Fourthly, higher inequality could impede the public legitimacy that environmental policy and politics need. Public legitimacy, in turn, is related to social stratification, with higher inequality being associated with lower public legitimacy for tough environmental legislation. An illustrative example of these patterns, which is related to several of the above points, is the outbreak of yellow vest riots in the streets of Paris in late 2018 (see Box 3.3).

Fifthly, at the international level, "the spatial and economic disjuncture between those who have largely caused and those who are most vulnerable to climate change has undermined the perceived need for global action" (Stoddard et al. 2021, p. 672). The spatial distance between the powerful rich and the powerless poor distorts the mechanisms of communication that would otherwise signal the need for an effective response.

Box 3.3 The yellow vests and the failure of the fuel tax increase

The yellow vest demonstrations were triggered by the French government's proposal to increase fuel taxes. The protesters were already frustrated with recent government reforms following the 2017 elections. A disproportionate burden of the government's tax reforms fell on the working class, particularly in rural areas, while the rich were able to enjoy cuts in wealth taxes and other privileges. Feelings of marginalization and public resentment over economic inequality predated these reforms, however, and France has a long tradition of contentious politics and protest movements (Grossman 2019, Mahfud and Adam-Troian 2021). There were diverse views within the movement, and some participants were concerned about climate change and were not necessarily opposed to carbon taxes as such (Mehleb et al. 2021).

The highly visible yellow vests served as a powerful anti-establishment symbol, as they represent hard outdoor labor. By November 17, 2018, nearly 300,000 yellow vest protesters were demonstrating across France (Grossman 2019). The street protests and roadblocks occasionally turned violent, resulting in 12 deaths and more than 3,000 injuries (among protesters, non-protesters, and the police), as well as the destruction of storefronts, cars, and public monuments (Mahfud and Adam-Troian 2021). Although participants were mainly from lower income groups, sympathizers came from all parts of France and abroad, and from all social classes, professions, and educational levels (Grossman 2019). The French government abandoned plans to raise the diesel tax by the end of 2018, which dampened the protests, though the movement continues.

This example shows that issues of structural inequality cannot be neglected when addressing policies for environmental change, and that the implementation of carbon taxation schemes requires more participatory, transparent, and equitable design (Mehleb et al. 2021).

Conclusion

The message of this chapter is that we cannot continue to neglect the social distribution of environmental problems in environmental debates and research. There is an urgent need for environmental social science to put uneven environmental effects at the center of environmental debates and analysis. We have presented some of the most promising and useful theories and concepts from environmental sociology. But why is this crucial?

First, equality is an end in itself. Equality makes for a better society, one that is pluralistic and heterogeneous, and where everyone has the opportunity to develop and flourish. Inequality is therefore a moral and a political problem. Various environmental studies, such as those focusing on intersectionality, environmental justice, ecological violence, and environmental disasters, have shown that various groups in society face very different options when it comes to suffering environmental impacts and enjoying environmental "goods." Such differences are generally not about choice; they are about social stratification beyond individual choice. The injustices are very manifest in the socio-material surroundings with local exploitation of natural resources and pollution, and they shape the material conditions of life (facet three) and are intimately linked to the operation of overarching institutions and power structures (facet five). Secondly, the topic is also an environmental problem in its own right. Inequality, as well as the broader issue of power asymmetries between various categories of actors (and intersections of categories such as gender, class, and ethnicity), plays an important role in causing environmental problems. It obstructs the work of mitigating environmental problems due to mechanisms such as relative poverty and collective dilemmas. Huge inequalities in a society creep under people's skin (Wilkinson and Pickett 2018); in other words, they have a huge impact on the inner lives of people and communities, and consequently block individual empowerment, environmental consciousness, and collective action (facets one and two). Accordingly, inequalities impede progressive politics, mobilization, movements, and work for societal transformation. The tendency toward polarization (e.g., in gender relations and urban-rural relations) and growing income disparities between and within countries must be broken if we are to succeed in working for transformative change. Increased equality is a crucial condition for transformative change.

We have discussed how society is stratified in multiple ways along categories such as social class, gender, ethnicity, age, income, and country of birth. Such differences can strongly influence the ways in which people suffer from environmental problems and disasters. We have also shown that social stratification contributes to our understanding of the human causes of environmental problems, adding to the repertoire of explanations that we focused on in Chapter 2. Social stratification and unequal exposure to environmental problems shape our different and sometimes highly polarized understandings of environmental problems, which is the subject of Chapter 4. Policies that are presented as solutions but fail to consider distributional aspects – the winners and losers of suggested reforms – will not have the potential to transform society (Chapter 5). Thus, the work toward transformative change requires that careful and serious consideration be given to the question of what will promote a more equitable society (Chapter 6).

Questions for reflection and discussion

What does the distribution of environmental "goods" and "bads" look like in the region where you live? Are some people there, or in a neighboring region, more vulnerable to environmental risks and hazards than others? Who are they in terms of social class, gender, ethnicity, or any other social categories? Are their concerns visible in media reporting? Are they represented in democratic decision-making? You might also go back and look at the power perspective introduced in Chapter 2. What additional insights might the three dimensions of power provide about the distribution of benefits and harms?

The chapter addressed problems of representation and democratic participation that are common to environmental issues with global, long-term, and irreversible consequences. Who, if anyone, is expressing the concerns of constituencies in other countries, or of future generations, or of non-human animals? What can they do to legitimately speak for the interests and needs of these groups?

What resources and positions in society enable a person to have sufficient motivation and capacity to engage in environmental advocacy, such as making environmentally conscious lifestyle choices, participating in environmental social movements, or demanding political action? Are these resources and positions blocked by circumstances related to social stratification?

The authors argue that the distributional effects of a disaster are related to the social circumstances before, during, and after the disaster. How can a community better prepare itself to meet the needs of all its residents in the event of a disaster?

The authors discuss how inequalities are growing within and between countries. How do you think growing inequality hinders efforts to improve the environment? What could be the most important measures to reduce the ecological and climate footprint of the most affluent?

Notes

1. In this book, we do not use the word "race," but we do speak of racial categories. Although the concept of race is commonly used in some countries and regions (e.g., the USA) and is used in society (with serious consequences), it is a biologically and geographically very questionable category. It is more relevant to speak of groups being racialized, which describes a situation in which a group occupying a privileged position (the norm) attributes certain characteristics to other groups of people: "To be racialized is to be the 'Other'" (Lockie 2018, p. 177). Structural racism exists when such attributions are systematic in society, with negative consequences for the racialized category. The term "ethnic groups" has a more cultural meaning than "race," and is accordingly less problematic, although we should not employ this term in a non-reflective way.
2. We are not claiming that news media cannot and do not cover slower, structural aspects of environmental destruction. Newspapers and broadcast news programs regularly produce in-depth analyses of disasters. Likewise, there is an increase in documentaries, which are better able to depict and explain slower forms of disasters and violence, such as the BBC's *Blue Planet II*, and Netflix's *Seaspiracy*. However, sudden disasters with immediate and dramatic consequences are the main types of disasters covered by the media.
3. This can also strengthen the identity and social bonds of a community hit by a disaster. The character of the post-disaster dynamics – the interaction and support that the affected community receives from external actors – has a significant impact on the community's identity and recovery (for cases of the breakdown and strengthening of community identity, respectively, see Edelstein 1988, Lidskog 2018).

References

Agyeman, J., Bullard, R. and Evans, B. (eds.) (2003) *Just sustainabilities: Development in an unequal world.* London: Earthscan.

Akenji, L., Bengtsson, M., Toivio, V., Lettenmeier, M., Fawcett, T., Parag, Y., Saheb, Y., Coote, A. et al. (2021) *1.5-degree lifestyles: Towards a fair consumption space for all.* Berlin: Hot or Cool Institute.

Albert, M.J. (2020) 'Capitalism and earth system governance: An ecological Marxist approach', *Global Environmental Politics*, 20(2), pp. 37–56. https://doi.org/10.1162/glep_a_00546

Anshelm, J. and Hultman, M. (2014) 'A green fatwā? Climate change as a threat to the masculinity of industrial modernity', *NORMA: International Journal for Masculinity Studies*, 9(2), pp. 84–96. https://doi.org/10.1080/18902138.2014.908627

Armstrong, C. (2019) *Why global justice matters: Moral progress in a divided world.* Cambridge: Polity.

Bacon, J.M. (2019) 'Settler colonialism as eco-social structure and the production of colonial ecological violence', *Environmental Sociology*, 5(1), pp. 59–69. https://doi.org/10.1080/23251042.2018.1474725

Beck, U. (1992) *Risk society: Towards a new modernity.* London: Sage.

Beck, U. (2016) *The metamorphosis of the world.* Cambridge, UK: Polity.

Berglez, P. and Lidskog, R. (2019) 'Foreign, domestic, and cultural factors in climate change reporting: Swedish media's coverage of wildfires in three continents', *Environmental Communication*, 13(3), pp. 381–394. https://doi.org/10.1080/17524032.2017.1397040

Boström, M. (2020) 'The social life of mass and excess consumption', *Environmental Sociology*, 6(3), pp. 268–278. https://doi.org/10.1080/23251042.2020.1755001

Boström, M. (2023) *The social life of unsustainable mass consumption.* Lanham: Lexington Books.

Boström, M. and Uggla, Y. (2016) 'A sociology of environmental representation', *Environmental Sociology*, 2(4), pp. 355–364.

Boström, M., Jönsson A.M., Lockie, S., Mol, A.P.J. and Oosterveer, P. (2015) 'Sustainable and responsible supply chain governance: Challenges and opportunities', *Journal of Cleaner Production*, 107, pp. 1–7. https://doi.org/10.1016/j.jclepro.2014.11.050

Boström, M., Uggla, Y. and Hansson, V. (2018) 'Environmental representatives: Whom, what, and how are they representing?' *Journal of Environmental Policy and Planning*, 20(1), pp. 114–127.

Brand, U. and Wissen, M. (2021) *The imperial mode of living: Everyday life and the ecological crisis of capitalism*. London: Verso Books.

Brown, P.R. and Zinn, J.O. (2022) *Covid-19 and the sociology of risk and uncertainty: Studies of social phenomena and social theory across 6 continents*: Cham: Palgrave Macmillan.

Brunsma, D.L., Overfelt, D. and Picou, J.S. (eds.) (2007) *The sociology of Katrina: Perspectives on a modern catastrophe*. Lanham, Md.: Rowman & Littlefield Publishers.

Bullard, R.D. (2000) *Dumping in Dixie: Race, class and environmental quality.* 3rd edn. Boulder, CO.: Westview.

Busca, D. and Lewis, N. (2015) 'The territorialization of environmental governance. Governing the environment based on just inequalities?', *Environmental Sociology*, 1(1), pp. 18–26. https://doi.org/10.1080/23251042.2015.1012617

Carfagna, L.B., Dubois, E.A., Fitzmaurice, C., Ouimette, M.Y., Schor, J.B., Willis, M. and Laidley, T. (2014) 'An emerging eco-habitus: The reconfiguration of high cultural capital practices among ethical consumers', *Journal of Consumer Culture*, 14(2), pp. 158–178. https://doi.org/10.1177/1469540514526227

Chancel, L. and Piketty, T. (2015) *Carbon and inequality: from Kyoto to Paris: Trends in the global inequality of carbon emissions (1998–2013) and prospects for an equitable adaptation fund.* Paris: Paris School of Economics.

Curran, D. (2013) 'Risk society and the distribution of bads: Theorizing class in the risk society', *The British Journal of Sociology*, 64(1), pp. 44–62. https://doi.org/10.1111/1468-4446.12004

Davidson, D.J. (2018) 'Metabolism' in M. Boström and D. Davidson (eds.) *Environment and society.* Cham: Palgrave Macmillan, pp. 47–70.

Davies, T. (2022) 'Slow violence and toxic geographies: "Out of sight" to whom?', *Environment and Planning C: Politics and Space*, 40(2), pp. 409–427. https://doi.org/10.1177/2399654419841063

DesJardins, J.R. (2006) *Environmental ethics: An introduction to environmental philosophy.* 4th edn. London: Thomson Learning/Wadsworth.

Dickens, P. (1996) *Reconstructing nature: Alienation, emancipation, and the division of labour*. London: Routledge.

Edelstein, M. (1988) *Contaminated communities: The social and psychological impacts of residential toxic exposure*. Boulder, CO: Westview Press.

Ergas, C., McKinney, L. and Bell, S.E. (2021) 'Intersectionality and the environment' in B.S. Caniglia et al. (eds.) *Handbook of environmental sociology.* Cham: Springer, pp. 15–34.

Erikson, K. (1995) *New species of trouble: Human experience of modern disasters*. New York: Norton.

Fischer, F. (2018) 'Environmental democracy: Participation, deliberation and citizenship' in M. Boström and D. Davidson (eds.) *Environment and society*. Cham: Palgrave Macmillan, pp. 257–280.

Fridell, G. (2019) 'Conceptualizing political consumerism as part of the global value chain' in M. Boström, M. Micheletti and P. Oosterveer (eds.) *The Oxford handbook of political consumerism.* Oxford University Press, pp. 249–274.

Gössling, S. (2019) 'Celebrities, air travel, and social norms', *Annals of Tourism Research*, 79, 102775. https://doi.org/10.1016/j.annals.2019.102775

Gössling, S. and Humpe, A. (2020) 'The global scale, distribution and growth of aviation: Implications for climate change', *Global Environmental Change*, 65, 102194. https://doi.org/10.1016/j.gloenvcha.2020.102194

Grossman, E. (2019) 'France's yellow vests – Symptom of a chronic disease', *Political Insight*, 10(1), pp. 30–34. https://doi.org/10.1177/2041905819838152

Gundelach, B. and Kalte, D. (2021) 'Explaining the reversed gender gap in political consumerism: Personality traits as significant mediators', *Swiss Political Science Review*, 27(1), pp. 41–60. https://doi.org/10.1111/spsr.12429

Henderson, J. (2020) 'EVs are not the answer: A mobility justice critique of electric vehicle transitions', *Annals of the American Association of Geographers*, 110(6), pp. 1993–2010. https://doi.org/10.1080/24694452.2020.1744422

Hoover, E. (2018) 'Environmental reproductive justice: Intersections in an American Indian community impacted by environmental contamination', *Environmental Sociology*, 4(1), pp. 8–21. https://doi.org/10.1080/23251042.2017.1381898

Huber, M.T. (2022) *Climate change as class war: Building socialism on a warming planet.* London: Verso.

Isaksen, K.J. and Roper, S. (2008) 'The impact of branding on low income adolescents: A vicious cycle', *Psychology & Marketing*, 25(11), pp. 1063–1087. https://doi.org/10.1002/mar.20254

Jorgenson, A.K., Fiske, S., Hubacek, K., Li, J., McGovern, T., Rick, T., Schor, J.B., Solecki, W., York, R. and Zycherman, A. (2019) 'Social science perspectives on drivers of and responses to global climate change', *WIREs Climate Change*, 10, e554. https://doi.org/10.1002/wcc.554

Kartha, S., Kemp-Benedict, E., Ghosh, E., Nazareth, A. and Gore, T. (2020) *The carbon inequality era: An assessment of the global distribution of consumption emissions among individuals from 1990 to 2015 and beyond.* Oxfam & Stockholm Environment Institute.

Kennedy, E.H. and Kmec, J. (2018) 'Reinterpreting the gender gap in household pro-environmental behaviour', *Environmental Sociology*, 4(3), pp. 299–310. https://doi.org/10.1080/23251042.2018.1436891

Kenner, D. (2019) *Carbon inequality: The role of the richest in climate change.* London: Earthscan.

King, L. (2020) 'Environmental justice and capitalism' in K. Legun, J. Keller, M. Carolan and M. Bell (eds.) *The Cambridge handbook of environmental sociology.* Cambridge: Cambridge University Press, pp. 452–469.

Koh, D. (2020) 'COVID-19 lockdowns throughout the world', *Occupational Medicine*, 70(5), p. 322. https://doi.org/10.1093/occmed/kqaa073

Kopnina, H. (2014) 'Environmental justice and biospheric egalitarianism: Reflecting on a normative-philosophical view of human-nature relationship', *Earth Perspectives*, 1(8), pp. 1–11. https://doi.org/10.1186/2194-6434-1-8

Krange, O., Kaltenborn, B.P. and Hultman, M. (2019) 'Cool dudes in Norway: Climate change denial among conservative Norwegian men', *Environmental Sociology*, 5(1), pp. 1–11. https://doi.org/10.1080/23251042.2018.1488516

Kruse, K.M. (2005) *White flight: Atlanta and the making of modern conservatism.* Princeton, N.J.: Princeton University Press.

Lidskog, R. (2018) 'Invented communities and social vulnerability: The local post-disaster dynamics of extreme environmental events', *Sustainability*, 10(12), pp. 44–57. https://doi.org/10.3390/su10124457

Lidskog, R. and Elander, I. (2007) 'Representation, participation or deliberation? Democratic responses to the environmental challenge', *Space and Polity*, 11(1), pp. 75–94. https://doi.org/10.1080/13562570701406634

Lidskog, R. and Sjödin, D. (2016) 'Extreme events and climate change. The post-disasters dynamics of forest fires and forest storms in Sweden', *Scandinavian Journal of Forest Research*, 31(2), pp. 148–155. https://doi.org/10.1080/02827581.2015.1113308

Lidskog, R. and Zinn, J. (2024) 'Reflexive modernization and risk society thesis' in C. Overdevest (ed.) *Edward Elgar encyclopedia of environmental sociology.* Cheltenham: Edward Elgar, pp. 471–476.

Lidskog, R., Elander, I. and Standring, A. (2020) 'COVID-19, the climate and transformative change: Comparing the social anatomies of crises and their regulatory responses', *Sustainability*, 12(16), p. 6337. https://doi.org/10.3390/su12166337

Lockie, S. (2018) 'Privilege and responsibility in environmental justice research', *Environmental Sociology*, 4(2), pp. 175–180. https://doi.org/10.1080/23251042.2018.1460936

Löwy, M. (2020) 'The Ecosocialist alternative', in K. Legun, J. Keller, M. Carolan and M. Bell (eds.) *The Cambridge handbook of environmental sociology*. Cambridge: Cambridge University Press, pp. 143–151.

Mahfud, Y. and Adam-Troian, J. (2021) '"Macron demission!": Loss of significance generates violent extremism for the Yellow Vests through feelings of anomia', *Group Processes & Intergroup Relations*, 24(1), pp. 108–124. https://doi.org/10.1177/1368430219880954

Malin, S.A. and Ryder, S.S. (2018) 'Developing deeply intersectional environmental justice scholarship', *Environmental Sociology*, 4(1), pp. 1–7. https://doi.org/10.1080/23251042.2018.1446711

Malin, S.A., Ryder, S. and Galvão L.M. (2019) 'Environmental justice and natural resource extraction: Intersections of power, equity and access', *Environmental Sociology*, 5(2), pp. 109–116. https://doi.org/10.1080/23251042.2019.1608420

Marks, D. (2018) 'The political ecology of uneven development and vulnerability to disasters' in R. Padawangi (ed.) *Routledge handbook of urbanization in Southeast Asia.* London: Routledge, pp. 345–354.

McCright, A.M. and Dunlap, R. (2011) 'Cool dudes: The denial of climate change among conservative white males in the United States', *Global Environmental Change*, 21(4), pp. 1163–1172. https://doi.org/10.1016/j.gloenvcha.2011.06.003

Medwinter, S.D. (2020) 'Reproducing poverty and inequality in disaster: Race, class social capital, NGOs, and urban space in New York City after superstorm Sandy', *Environmental Sociology*, 7(1), pp. 1–11. https://doi.org/10.1080/23251042.2020.1809054

Mehleb, R.I., Kallis, G. and Zografos, C. (2021) 'A discourse-analysis of yellow vest resistance against carbon taxes', *Environmental Innovation and Societal Transition*, 40, pp. 382–394. https://doi.org/10.1016/j.eist.2021.08.005

Merton, R. (1938) 'Social structure and anomie', *American Sociological Review*, 3(5), pp. 672–682. https://doi.org/10.2307/2084686

Micheletti, M. (2004) 'Why more women? Issues of gender and political consumerism' in M. Micheletti, A. Follesdal and D. Stolle (eds.) *Politics, products, and markets: Exploring political consumerism past and present.* New Brunswick, London: Transaction Publishers, pp. 245–264.

Mythen, G. (2005) 'Employment, individualization and insecurity: Rethinking the risk society perspective', *The Sociological Review*, 53(1), pp. 129–149. https://doi.org/10.1111/j.1467-954X.2005.00506.x

Nixon, R. (2011) *Slow violence and the environmentalism of the poor.* Cambridge, Mass.: Harvard University Press.

Olofsson, A., Öhman, S. and Giritli Nygren, K. (2016) 'An intersectional risk approach for environmental sociology', *Environmental Sociology*, 2(4), pp. 346–354. https://doi.org/10.1080/23251042.2016.1246086

Otto, I.M. et al. (2020) 'Social tipping dynamics for stabilizing earth's Climate by 2050', *PNAS*, 117(5), pp. 2354–2365. https://doi.org/10.1073/pnas.1900577117

Oxfam (2020) *Confronting carbon inequality. Putting climate justice at the heart of the COVID-19 recovery.* Oxfam Media Briefing. https://www.oxfam.org/en/research/confronting-carbon-inequality

Pellow, D.N. (2020) 'Expanding critical and radical approaches to environmental justice' in K. Legun, J. Keller, M. Carolan and M. Bell (eds.) *The Cambridge handbook of environmental sociology*. Cambridge: Cambridge University Press, pp. 399–415.

Roberts, T.J., Pellow, D. and Mohai, P. (2018) 'Environmental justice' in M. Boström and D. Davidson (eds.) *Environment and society: Concepts and challenges.* Palgrave Macmillan, pp. 233–256.

Sayer, R.A. (2015) *Why we can't afford the rich.* Bristol: Policy Press.

Seamster, L. and Purifoy, D. (2021) 'What is environmental racism for? Place-based harm and relational development', *Environmental Sociology*, 7(2), pp. 110–121. https://doi.org/10.1080/23251042.2020.1790331

Stoddard, I., Anderson, K., Capstick, S., Carton, W., Depledge, J., Facer, K., Gough, C., Hache, F. et al. (2021) 'Three decades of climate mitigation: Why haven't we bent the global emissions curve?', *Annual Review of Environment and Resources*, 46, pp. 653–689. https://doi.org/10.1146/annurev-environ-012220-011104

Tierney, K. (2012) 'A bridge to somewhere: William Freudenburg, environmental sociology, and disaster research', *Journal of Environmental Studies and Sciences*, 2, pp. 58–68. https://doi.org/10.1007/s13412-011-0053-9

UN (1987) *Our common future. Report of the world commission on environment and development.* Nairobi: United Nations Environment Programme.

Wallerstein, I.M. (2004) *World-systems analysis: An introduction*. Durham: Duke University Press.

Wiedmann, T., Lenzen, M., Keyßer, L.T., and Steinberger, J.K. (2020) 'Scientists' warning on affluence', *Nature Communications*, 11(1), 3107. https://doi.org/10.1038/s41467-020-16941-y

Wilkinson, R. and Pickett, K. (2011) *The spirit level: Why greater equality makes societies stronger.* London: Bloomsbury.

Wilkinson, R. and Pickett, K. (2018) *The inner level: How more equal societies reduce stress, restore sanity and improve everybody's well-being.* London: Allen Lane.

World Bank (2023) 'Poverty'. https://www.worldbank.org/en/topic/poverty/overview (accessed 26 June 2023).

World Food Program (2023) 'A global food crisis'. https://www.wfp.org/global-hunger-crisis (accessed 26 June 2023).

Wright, E.O. (1984) 'A general framework for the analysis of class structure', *Politics & Society*, 13(4), pp. 383–423. https://doi.org/10.1177/003232928401300402

4 Understandings

The social sense-making of environmental problems

Climate change, biodiversity loss, animal extinction, urban air pollution, and exceeding planetary boundaries. Greenhouse gases, pollutants, and nuclear waste. These and other phrases are circulating in society – in the daily news, social media, and political debates. They signal that our Earth is under threat and that we need to do something soon. Otherwise places will become uninhabitable and, in the worst case, our civilization will collapse.

Other stories are also circulating, stories of economic growth, human prosperity, human rights, and democratic development. There are also stories about how society is on the right track, promising that current environmental problems – like past ones such as acidification and ozone-layer depletion – will be solved through technology, planning, and policy. Environmentally harmful substances are being phased out, dirty technologies are being replaced by smarter and greener ones, environmental taxes and economic incentives are increasingly being utilized to make it expensive to pollute, and environmental awareness is spreading, leading people to choose environmentally friendly products and to place environmental demands on producers and politicians.

A third and very different type of story is also flourishing, one that includes phrases such as fake news, global elites, bought science, and conspiracies. These are stories that do not tell us whether we are on the right or wrong environmental track, but instead claim we are being deceived by global political and business elites with a hidden agenda, who use the vocabulary of environmental crisis to manipulate us into uncritically submitting to stricter regulations that limit our freedom.

These three stories illustrate the importance of how we make sense (or nonsense) of environmental problems. We might wonder how they can be so different. An environmental problem can often be interpreted in very different ways, with significant implications for how much we will prioritize it and what actions we will take in response. The struggle over the environment is thus also a struggle over our understanding. Are we facing a crisis or not? Is economic growth part of the problem or part of the solution? Will (new) technology create a sustainable society or are structural changes necessary? Do we need to question and change our way of life?

In this chapter we will explore how we make sense of environmental problems, how environmental problems are defined and narrated, and how people and organizations develop claims about environmental problems. This concerns all facets

DOI: 10.4324/9781032628189-4

of the social, from the inner lives of people and communities (emotions, values, knowledge, identities), social relations and interactions (how understandings relate to issues of belonging, dialogue, etc.), social practices and technological devices such as carbon accounting, how our sense-making relates to socially stratified experiences, and institutional dimensions, such as the role of science in society. We start from the assumption that our understanding of environmental issues is a much broader social phenomenon than just a compilation of scientific facts, and also relates to familiar knowledge, norms, values, and emotions. We stress that all forms of knowledge, including scientific ones, are socially embedded and must be understood as such.

In line with this argument, we stress that science is a necessary but not sufficient factor in solving environmental problems. To motivate people to act, scientific knowledge must make sense to actors. We will therefore present two important social science concepts that help to explain how societies and societal actors make sense of environmental problems, namely storylines and frames. We will then describe three frames – technological, cognitive, and structural fixes – which are frequently used by actors when proposing environmental solutions. However, these are often applied in an overly narrow sense that fails to take social aspects into account.

We will then turn our attention to the question of why scientific facts can be assessed very differently by different people. We show that they can be used to support stories of both environmental progress and environmental decline. This does not imply a relativistic view of science, however, but only that it is the wider interpretation of scientific results that motivates action or inaction. We will also look at the role of environmental expertise. It is crucial to reflect on who qualifies as an environmental expert, which includes processes of social legitimation. Moreover, in everyday life and politics, expert knowledge interacts with other ways of knowing and understanding the world, including practical knowledge. The persistent phenomenon of climate denialism is a related theme, which we address below. Indeed, we live in a time when attempts to remedy environmental problems threaten fundamental interests in society, and there is constant greenwashing and production of counterclaims. Finally we discuss the key insights of this chapter by relating them to the five facets of the social.

From fact-finding to sense-making

Scientific knowledge is crucial to understanding environmental problems. The vast majority of calls for environmental action – whether by governments, corporations, environmental movements, or individuals – contain direct or indirect references to scientific knowledge. One reason for this is that many environmental problems are increasingly diffuse and beyond direct perception (Beck 1992). In some cases, the consequences are delayed in time; the results of actions taken today will affect us or unborn generations in the future. In other cases, the consequences are spatially distributed. The regional distribution of airborne pollutants, for example, can make it difficult for people to make a connection between cause and effect. Driving an electric car means less climate impact (in terms of carbon dioxide emissions) and better urban air quality (e.g., in terms of sulfur and nitrogen emissions), but the production of the car has itself caused environmental problems. For example, the battery contains lithium and cobalt, which were mined in an environmentally detrimental way and with poor working conditions for the miners. These environmental impacts may go unnoticed by the car owner, who only looks at the environmental facts supplied by the manufacturer. Some of these environmental impacts are direct and noticeable to those exposed to them, but "invisible" to those who use the products at a distance from their production. Other environmental impacts are diffused, spread in space and time, are therefore not directly perceptible, and have complex links between causes and effects. In any case, science is critical to raising awareness of environmental problems.

Scientific research is highly specialized, however, and to make it relevant and accessible to society more broadly, international expert organizations have emerged that are tasked with synthesizing, translating, and communicating scientific knowledge to decision makers, stakeholders, and the public (Beck et al. 2014, Lidskog and Sundqvist 2015). The Intergovernmental Panel on Climate Change (IPCC) is the best-known of these organizations, but similar expert organizations exist for many environmental issues (Jabbour and Flachsland 2017).

As environmental sociologists often point out, we live in a society where many environmental risks are fundamentally research-dependent, and where scientific experts play a central role not only in guiding policy decisions, but also in guiding people's everyday life choices. It seems that when it comes to leading a healthy life on a healthy planet, people have no choice but to trust scientific experts, and governments may wonder why people do not always follow their recommendations (and conversely, people may wonder why governments do not listen to scientific advice). Why do people ignore some environmental risks when there is strong scientific evidence that they are real threats to health and the environment? This problem is often formulated in terms of public ignorance of science, with the solution being more and better scientific communication (Wynne 1992).

We share the view that science is a necessary part of solving environmental problems, but we argue that it is not sufficient in itself. As we stressed in Chapter 2, people are socially embedded, which means that problems – even environmental ones – look different depending on who you are and where you live. Scientific knowledge must fit into, or at least relate to, the ways in which people and organizations understand and make sense of the world; otherwise it may seem irrelevant or not even be noticed. It is not enough to scientifically characterize the causes, effects, and remedies of an environmental problem, and to disseminate this knowledge in society. If that were the case, the simple solution to the environmental crisis would be more and better-communicated science. But this view is mistaken and tends to be part of the problem rather than part of the solution. An overly strong belief in the importance of science conceals the social side of the question, namely that science is only one of many domains that influence people, organizations, and society as they develop actions. A prerequisite for science to have an effect is that it is not only communicated and understood, but also meaningful – that it appears relevant and credible to actors in various contexts. Thus, instead of stressing the need to educate the public (in scientific literacy), there is a need to educate scientists (and governments) about how people appropriate messages and what motivates action (Wynne 1992).

Furthermore, those who ask why people do not listen to science also need to raise the question why governments do not do so either. Placing these questions on an equal footing makes it harder to embrace the naive view that more and better-communicated environmental knowledge is sufficient to halt environmental destruction. Instead, aspects of power and special interests must be included in an analysis of why environmentally detrimental activities are maintained and often even strongly supported. For example, state subsidies for fossil fuel consumption are estimated to currently amount to more than 400 billion USD per year (IEA 2019). This is despite the Paris Agreement and governments' talk about the importance of combating climate change.

It should also be noted that science is not only insufficient for promoting the environmental cause, but can sometimes be an unreliable friend of it (Yearley 1991). This is because science itself has contributed to many environmental problems, both historically and currently. This is stressed by macrosocial environmental theories. For instance, the theories of the treadmill of production, risk society, ecological modernization, and postcolonialism (presented in Chapter 2) all view science (and technology) as an integral part of society, integrated into production and consumption and used by many actors to further their interests. Thus, the common idea that "science speaks truth to (political) power" is more of an ideal than a reality, because scientific claims can be wrong, and scientists can ally themselves with interest groups. This makes it all the more important to foster a broad and critical discussion about scientific claims and science's role in society (Lidskog et al. 2022). It should be borne in mind that such a critical discussion of science is something completely different from conspiracy theories and claims about alternative facts. As Robert Merton (1910–2003), one of the founders of the sociology of science, pointed out, science is governed by norms, one of which is organized skepticism. This means that scientific knowledge claims must not be uncritically accepted, but rather need to be subjected to rigorous testing and critical evaluation (Merton 1942/1973). Thus, all knowledge claims need to be critically examined, even scientific ones.

Why knowledge isn't enough

There is a widespread belief in society that more knowledge will solve environmental (and other) problems. The call to "listen to the science," promoted by Greta Thunberg and the international climate movement *Fridays for Future*, among others, is often seen as imperative for solving the environmental crisis (Thunberg 2019). Scientists themselves are also making great efforts to disseminate scientific knowledge about the seriousness of the planetary situation, and believe or hope that this will persuade political and economic elites to join the public in taking action.

Numerous social scientific studies, however, show that people are not guided by knowledge alone, and that knowledge in itself is not sufficient to trigger action (Heberlein 2012, Shove 2003). This is the case even when scientific knowledge is well communicated and properly understood. One reason for this is that in order to guide action, knowledge must be combined with values and norms, as well as with other forms of knowledge, such as practical knowledge. Simply knowing something gives little indication of the best way to respond. Informing the public that today's consumption patterns will adversely affect future generations, or that they are already affecting people in other places, will not necessarily lead to any action. The message must be linked to deeply held values that it is wrong to live in ways that limit the ability of others to live, including future generations.

Furthermore, norms may say that it is wrong to act in a certain way, or that it is wrong not to act at all. Nevertheless, a person, group or organization may fail to take action when they know they should. This is because it is also necessary to combine knowledge with emotions (Barbalet 2002, Engdahl and Lidskog 2014, Finucane 2013). If a message about the need to act does not evoke any emotions, it is unlikely that action will be taken, despite shared agreement about the situation and what needs to be done. Thus, to trigger action, it is not enough to explain the world (factual knowledge) and describe what actions are needed (normative orientation and connection to values). It is also essential to prioritize the issue and create commitment, which is done by connecting the issue to people's emotions (emotional appeals) (Lidskog et al. 2020). We can also ask what kinds of emotions best motivate action, and many social psychologists would say that these cannot be exclusively negative emotions. Harsh warnings or images of environmental destruction may evoke fear, anxiety, frustration, guilt, and anger. Such emotions can lead to passivity, but it is also possible for them to be forces of "critical emotional awareness" (Ojala 2016), and when they are combined with more positive emotions, such as hope, enthusiasm, and pride in doing the right thing, they can become creative forces for action.

The need for knowledge to be meaningful and relevant also applies to triggering action by organizations. As described in Chapter 2, organizations act according to their own norms and goals. The organizational theorist Karl E. Weick has stressed that organizations attempt to make sense of their external environment, which includes, for example, legislation, cultural trends, the actions of other organizations, and the natural environment (Weick 1995, Weick et al. 2009). An organization needs to gather and process information – to constantly ask the question "what's going on here?" – in order to navigate its organizational environment, which is often very complex and uncertain. Sometimes new information can easily be interpreted in line with the organization's original understanding. Other times, new information does not fit with its original understanding, and it must develop a sense of what it is up against. By noticing, bracketing, and labeling what is going on, an organization develops a plausible and coherent picture of the situation that facilitates certain options and actions and constrains others. This is especially true when organizations face unexpected events, such as bank failures, natural and technological disasters, and conflicts (Weick and Sutcliffe 2015). By making sense of their organizational environment, organizations can learn from experience and change their practices.

In sum, the belief that more and better distributed environmental knowledge is a way forward is in most cases a dead end; knowledge mainly facilitates action when already highly motivated people and organizations request it to guide their action. Nor is the mere appeal to norms and emotions sufficient to trigger action. Even when people and organizations are well informed about the environment, embrace environmental values, and are socially embedded in settings with strong norms about saving the planet, they may nevertheless continue to engage in environmentally harmful actions. This is often referred to as the value-action gap or the attitude-behavior gap (see Chapter 2 on microsocial explanations and Chapter 5 on the gap between statements and action).

Storylines and frames: Making environmental issues meaningful

No one is motivated by scientific facts in isolation. It is their wider interpretation that makes them meaningful. Within a particular scientific community, that meaning may be self-evident. A climate scientist understands what is meant by radiative forcing capacity (RF) and a plant physiologist knows what is meant by the exposure index AOT40, but most people do not. There are other concepts and facts that make sense to us, because we have tacitly and gradually acquired broader understandings that make facts meaningful. In the 1980s, most car buyers would have been confused if information about a car model had included how much carbon dioxide it emitted per kilometer. "Why is this information being provided? Carbon dioxide is something that all living things exhale, and it is not toxic or harmful to the environment." Today, most people associate carbon dioxide with global warming, so they understand why such information is included.

It's the larger story that matters

We are constantly, and often unconsciously, interpreting and making sense of facts, figures, and events. When science is claimed to be a crucial factor for environmental awareness and policy, it is not science as a collection of individual facts that is meant, but science as a wider scientific narrative about the state of the environment and possible environmental futures. The larger story is what matters and is what makes individual facts meaningful. When scientific concepts, measures, and figures are translated into terms that make sense to ordinary people, they must be part of a wider story. For example, they can tell a story of progress or decline. They can be part of a story about how industrial enterprises exploit people, places, and the environment – or how they now operate with great environmental sensitivity and in dialogue with local residents. They can be part of a story about how people are becoming more environmentally aware and are making green choices in their daily lives – or about how they have become green only in words, while in their deeds they continue to make choices that are anything but ecologically sustainable.

Typically, a *storyline* gives a historical account of a problem, its causes, and its consequences. Storylines motivate, guide, and legitimize decisions and actions. By saying what has happened, why, and what to do about it (explicitly or implicitly), storytelling combines factual statements with a normative orientation to facilitate action (Lidskog et al. 2020, Rosenbloom 2018). Successful storytelling not only explains the world in a trustworthy way but also motivates action. A central means of doing this is to engage the listener emotionally (Arnold 2018). As has long been stressed in the field of rhetoric, to teach and engage listeners, an orator must use not only logos (knowledge) and ethos (trustworthiness), but also pathos (appeals to emotions). Visualization is often a very important factor in evoking emotions (Boholm 2015, Hansen and Machin 2013). Representing the consequences of abstract and uncertain future risks in images gives them concrete form and content. Climate change can be illustrated by starving polar bears, koalas unable to escape a forest fire, a dejected farmer standing in a parched cornfield, or a community destroyed by a flood. Other ways of representing abstract risks are to use symbols and analogies, or to present them in the form of indices and graphs that summarize complex and broad processes of change (Stone 2012). Other senses are also involved. Ecological claims about organic food are often connected with stories about freshness and flavor. Such storylines and broader understandings are not only developed by scientists and expert organizations, and are not necessarily based on a scientific understanding of an issue. Storytelling is a fundamental element of all cultures, and humanities scholars have analyzed the role, construction, and structure of stories and what makes them meaningful and persuasive (Ricœur 1995).

Closely related to storylines is the concept of *frames*. Frames and framing processes are used to explore how meaning is ascribed to different phenomena. Originally coined by Erving Goffman (introduced in Chapter 2), the term "frames" describes how meaning is created, maintained, and negotiated in microsocial interactions (Goffman 1974/1986). This idea has been further developed to be applicable to understanding sense-making processes in different settings. Frames are structures of beliefs, values, and perceptions through which actors conceptualize, interpret, and understand an issue (Schön and Rein 1994). Through frames, a complex issue or situation is translated into simpler terms, and certain aspects of it are emphasized. By reducing the complexity of an issue to highlight a specific understanding of what it fundamentally concerns, frames tell what is at stake and mobilize opinion for particular kinds of action (or inaction). Whereas a storyline provides a story about a problem, a condensed narrative about an issue – such as why an environmental disaster occurred or why a particular environmental problem has developed – frames are more general and need not necessarily be linked to a specific problem or problem trajectory. Frame analysis can be useful for analyzing debates within politics and policy-making, as we show in Box 4.1, and it has also been widely used to analyze the role of the media in framing environmental messages. (On the role of the media in environmental policy and public debates, see Campbell 2021, Hannigan 2020, 2022, and Lester 2010.)

Box 4.1 Frame conflicts: Policy disagreement or policy controversy?

Much of politics and policymaking consists of struggles over the dissemination of particular frames (Schön and Rein 1994). If a particular frame is successful, it will be agreed upon and taken for granted by the relevant actors, and no conflicts will occur. Disagreements may arise between actors despite their sharing a frame. In these cases – referred to as *policy disagreements* – the actors understand a situation in a similar way, and therefore the conflict can be resolved by examining the facts or gaining more knowledge about the situation. In many cases, however, actors hold different frames. This constitutes the basis of a *policy controversy*. Such controversies cannot be resolved simply by appealing to facts, because the actors attach different degrees of importance to the facts, or interpret the same facts differently. In line with the frames they adhere to, actors ascribe certain characteristics to an issue, such as hazardous or harmless, safe or risky, predictable or unpredictable, natural or unnatural, important or unimportant.

For example, nuclear power is seen by some actors as the most environmentally friendly option for producing electricity, because it does so without emitting carbon dioxide. Within this camp, there may be policy disagreements on issues such as the design of safety systems and the lifespan of various technological components, but these can be resolved by gaining more knowledge. Other actors stress that nuclear power is extremely risky – often referring to the nuclear meltdowns at Chernobyl (1986) and Fukushima (2011) – and claim that it is impossible to safely dispose of the highly radioactive nuclear waste for hundreds of thousands of years, or to safely extract necessary resources such as uranium. Here we have a policy controversy, because the diverging interpretations of nuclear power cannot be reconciled by gaining more knowledge about the risks and dangers, but only by changing the broader frames that guide their assessment. Thus, scientific facts are not unimportant, but what matters most is how they are interpreted – how they are framed and made part of a larger story.

The importance of storylines and frames can be exemplified by the concept of sustainable development, which has its roots in collaboration between international civil society, science, and government bodies. The idea stems from the report *World Conservation Strategy* (IUCN 1980), written by the International Union for Conservation of Nature and Natural Resources in collaboration with the World Wildlife Fund (WWF) and the United Nations Environment Programme (UNEP). This report was a platform for the UN Commission on Sustainable Development. Since its broad policy introduction through the UN Commission's report on environment and development (UN 1987), the concept has successfully marched through the political corridors and wider society, and is now embraced by almost all sectors and organizations. It has been widely endorsed by policy makers, business representatives, and interest groups, and it is extremely difficult to find an organization that does not openly subscribe to the notion of sustainable development or sustainability. In this sense, it has been extremely successful. Box 4.2 discusses some of the possibilities that the concept offers. However, a closer look reveals that it can be misused and made to mean almost anything, and can even be used to market oil extraction, long-haul flights, and gas-guzzling 300-horsepower car engines. Sustainability means very different things to different organizations and people, and how much it matters to an organization and in what way it is an empirical question. This shows that concepts and goals, even those that have been intensely negotiated and discussed before being agreed upon, are part of larger stories and are given different meanings and different priorities.

Box 4.2 Using the sustainability framing in storylines

The ideas of "modernization" and "progress" are salient features of the concept of "sustainable development," as it was formulated by the Brundtland Commission (UN 1987). With the growing recognition of ongoing environmental destruction, social conflicts, and even the risk of civilizational collapse, one might ask what role this concept can play. There are frequent accusations of greenwashing – the presentation of environmentally harmful activities as "sustainable." At the same time, there are a number of arguments for continuing to use the concept in storylines:

- It draws attention to the concept of *justice*, including both existing, living generations around the globe (intra-generational justice) and future generations (inter-generational justice) (see Chapter 3). At the same time, it makes overly anthropocentric assumptions.
- By focusing attention on *the future*, it serves as an important reminder to avoid short-termism in perspective and practice. For example, it is associated with important principles such as the precautionary principle and consideration of the needs of future generations.
- It emphasizes the need for *integration* of concepts that have historically been differentiated/separated, such as the well-known tripartite division of the "sociocultural," "economic," and "ecological" dimensions. While this may appear a bit simplistic, the sustainability discourse can help draw attention to the need for their integration (or analysis of their "synergies" and "trade-offs"). (See Boström 2012 for a discussion of the pros and cons of this conceptual distinction.)
- This search for integration also concerns the role of *knowledge*. The sustainability discourse advocates the need for better knowledge exchange and knowledge integration, both between different scientific disciplines and between science and other forms of knowledge.
- The concept of sustainability makes it possible to talk about the opposite, that something is *unsustainable*. This includes a critique that acknowledges that many activities done in the name of sustainability only serve to sustain the unsustainable (Blühdorn 2007, 2011).

In a similar way, social transformation can have very different meanings. It raises questions about what should be transformed, why something should be transformed, how it should be transformed, and by whom. These very far-reaching questions have many possible answers. In their book *Sustainability Transformations*, Linnér and Wibeck (2019) discuss how transformation is understood across societies. Employing the idea of sense-making processes, they find that different interests and values within and between societies explain why sustainability transformation means so many different things. Knowing how actors make sense of social transformation is central to understanding how social transformation can be initiated and governed. In Chapter 6, we further discuss the various meanings of transformation.

Three common frames: Technological, cognitive, and structural fixes

In this section we will discuss three common frames for handling an environmental problem: technological, cognitive, and structural fixes (Heberlein 2012, pp. 3–10), which are linked to different, but inadequate, proposals for solving environmental problems (see also Chapter 5).

A *technological fix* is the invention and application of technical measures to handle a problem and thereby avoid needing to change human behaviors and social practices. An example is how ozone-depleting substances were phased out and replaced by substitutes, thereby handling the problem of stratospheric ozone depletion without requiring the public to change its behavior (see Box 4.3). A current example is the development of renewable energy combined with the electrification of transport systems, which is expected to radically reduce climate emissions from road transport. If successful, this will mean that people will not need to reduce their use of cars. A more extreme example is the proposal to develop and deploy large-scale negative emission technologies (NETs) to remove greenhouse gases from the atmosphere. Carbon capture and storage, direct air capture and storage, ocean fertilization, and enhanced weathering have all been proposed as possible future methods to combat climate change (EASAC 2018, Hansson et al. 2021). Admittedly they are presented as part of a broader strategy for combating climate change (which also includes transformative changes), and our point is just that NETs are an example of a technological fix that reduces greenhouse gases without requiring any changes in human behavior.

Box 4.3 Beyond the technological fix: The case of stratospheric ozone depletion

Of course, some environmental problems can be solved without transforming society. There are plenty of historical cases in which chemicals that were found to be environmentally hazardous were replaced by less harmful ones. We currently see hope that this will be possible for fossil fuels – that biofuels and batteries can become the primary energy sources for transportation (with electricity being generated by solar, hydroelectric, or wind power). It should be noted, however, that a technological fix requires major social investments and often substantial changes in regulatory frameworks. An example of this is the measures taken to stop the depletion of stratospheric ozone (for a detailed discussion of this problem, see Grundmann 2001, Litfin 1994).

In the 1970s, research found that a class of substances called CFCs (chlorofluorocarbons) were depleting the stratospheric ozone layer. Through the development and use of other chemicals, it became possible to replace CFCs with other, less environmentally harmful, substances. But if we look into what happened, we see that even with technological fixes of this kind – solutions that seem to be purely based on technical innovation and not to need social change – society is still significantly involved. It was scientists and environmental organizations that raised the ozone issue and created awareness about it. Politicians were pressured to take action, and complicated international negotiations began. At the same time, research and development efforts were intensified to develop suitable chemical substitutes for CFCs. After many years of negotiations, an international convention was drafted, a plan for the phasing out of CFCs was established, and an international economic fund was created to facilitate the industrial transition. No major societal change was required, as society continued to manufacture and use refrigerators, aerosol cans, and circuit boards, the only difference being that CFCs were not used in these products. However, a major effort was demanded of various sectors and actors in society. Thus, society is also very active in solving seemingly "simple" environmental problems, where the solution is "only" to phase out an environmentally detrimental substance or to invent a technical artifact.

A *cognitive fix* involves providing sufficient information to the public to make people change their behavior and habits. Warning signs are based on this idea. We constantly encounter signs like "Don't leave valuables in the car," "Mind the gap," "Road work ahead." In the fight against the pandemic, there are messages and warnings such as "Please keep your distance," "Face coverings must be worn," and "Stop the spread, wash your hands." These are often powerful reminders of social norms and appropriate behavior. We might take our wallets and handbags with us from the car, take a big step when getting off the subway, drive more slowly, consider not standing too close to others in the grocery store, and wash our hands more often. In the environmental field, many information campaigns have been launched to get people to live in a more ecologically sustainable way by saving energy, sorting waste, and choosing more climate-friendly means of transportation. However, many of these campaigns have had limited impact, and as discussed above and in Chapter 2, increased knowledge alone will rarely lead to behavioral change. Knowledge needs to be linked to values, norms, and emotions to have more far-reaching and lasting effects (see also our discussion of transformative learning in Chapter 6). Thus, changing behavior through information is a simple and often chosen strategy, but it only works if the target groups are already motivated to take action.

A *structural fix* has to do with changing the context in which human behavior takes place. In this way, human behavior is changed not in conflict with people's attitudes and values, but in agreement with them. One example of a structural fix is "nudging," which means making it easier for people to make the right choices, such as by promoting greener standard options, encouraging people to use smaller plates to reduce food waste, or creating easily accessible and safe bike lanes (Thaler and Sunstein 2021). The idea is to shape people's behavior without requiring them to think about it. Another example is the use of pricing mechanisms. People and businesses may be reluctant to reduce their car travel, but if the government increases the gas tax it will become more costly, and people and businesses may reduce their driving or choose other modes of transportation. For economic reasons, people may commute to work by public transportation or by bike, rather than by car. It is important to note that this strategy may have unintended consequences. As discussed in Chapter 3, one effect of the stratification of society is that people have different kinds and amounts of resources. For some social segments, an increased gas tax does not have an impact, because it is still affordable for them to take the car, and they will continue with their driving practices. For other, less affluent segments, the higher gas prices may have far-reaching consequences for their daily lives. Thus, pricing mechanisms may not only reproduce inequalities in society, but even increase them. This also explains why some structural fixes are hard to implement. For example, the increase in fuel taxes in France in 2018 sparked the yellow vest movement because it disproportionately affected the working and middle classes – at the same time as the government had just abolished the wealth tax. What was initially framed as an environmental reform soon changed into a question of social justice (see Box 3.3).

Frames associated with technological, cognitive, and structural fixes are all common and relevant ways of dealing with environmental issues. Governments make use of all of them, to varying degrees, to address particular environmental problems, and they are often combined. For example, climate change is addressed simultaneously with the development of renewable energy, the dissemination of environmental knowledge, and the creation of economic incentives. Our point here is that all these fixes, whether technological, cognitive, or structural, are based on narrow understandings and framings of environmental problems and how society works. They all tend to rely on an overly positive view of market forces and lack a more fundamental critique of existing societal structures and cultural patterns that reproduce the problems. As we discuss in Chapter 5, if they are applied too narrowly, they can become part of the problem rather than part of the solution.

In his book *How to Avoid a Climate Disaster: The Solutions We Have and the Breakthroughs We Need* (2021), Bill Gates paints a very bleak picture of the climate crisis and argues that unless we take urgent and decisive action, we will soon face a climate catastrophe. He notes that it will be difficult to achieve zero greenhouse gas emissions by 2050. But this message is followed by the good news that it is still achievable, and that it can be done by deploying existing technological tools (such as solar and wind power) faster and smarter, and by developing breakthrough technologies and deploying them on a large scale globally. This is an example of a technological fix that requires government decisions and public funding to implement, but does not demand any radical changes in people's consumption and social practices. The book has much in common with the theory of ecological modernization (presented in Chapter 2), and its belief that society can ecologically modernize itself into a condition of sustainability. Bill Gates's understanding of climate change and its solution is based on a particular framing of the climate problem that specifies what should be fixed, why it needs to be fixed, how it should be fixed, and who should be responsible for fixing it.

Gates recognizes that there is a growing global climate movement and there are lofty political goals for solving the climate change. What is needed now is to develop a concrete plan and take action to achieve these goals, and this plan is based primarily on developing zero and negative emission technologies. This in turn requires large investments in research and innovation, as well as in technical infrastructure (e.g., electricity distribution). It is also necessary to develop economic incentives to make new technologies economically competitive (e.g., a carbon dioxide tax). The public also has a role to play, but the main share of our climate emissions comes from the larger technical systems of which we are a part, and therefore it is these systems that need to change. In this framing, the public's role is mainly to exert political pressure and to help increase the demand for climate-friendly products (e.g., by buying an electric car, using green electricity, or eating meat alternatives a couple of times a week). There is no advice to reduce car and air travel or to reduce consumption; instead, the solution is to develop climate-friendly alternatives. Economic growth is seen as crucial; the world needs more goods and services, and this will be made possible through technological innovations that can create green energy, transportation, agriculture, shipping, and heating/cooling.

Roy Scranton tells a very different story. His award-winning book *Learning to Die in the Anthropocene: Reflections on the End of a Civilization* (2015) agrees with Bill Gates's diagnosis: that society is currently on the wrong track. Our earlier transition to a society based on fossil energy is now leading to a global climate catastrophe. But the way forward that Scranton presents is fundamentally different from that of Gates. For Scranton, it is naive to believe in fixes such as technological innovations, green taxes, or global international agreements, because none of these address the root causes: the structure of present-day society and the lifestyle it generates. There is no external enemy to blame, but instead we are our own worst enemy (even if some actors are definitely more powerful and responsible for the current situation than others). All of us – states, organizations, people – are so dependent on fossil energy flows that there is no future for this society, and thus no solutions can be found within this society. Scranton therefore argues that our civilization (like the Roman one) must die, along with its values and privileges. What we need to do now is to prepare for this by thinking about what humanity should preserve and bring to a future civilization.

The different stories told by Bill Gates and Roy Scranton show how scientific facts, relevant policies, and key actors are all part of broader interpretative frames and storylines. This explains why so many different solutions to the climate crisis have been proposed. Our intention in this chapter is not to provide a yardstick that tells us whose interpretation is the correct one, and what measures will be most effective in combating climate change. What we do argue is that all policies, plans, and solutions must take into account the underlying social reality. If a proposed solution does not consider this key aspect – how people, organizations, and society work – it will not be successful. It may have limited effects, or it may have its intended effect but also have severe unintended and unanticipated effects (as the yellow vest movement shows). Thus, any fixes must take into account the five facets of the social, and even if it is not possible to know in advance what a policy, decision, or implemented measure will lead to, the inclusion of more and broader knowledge will allow for more effective action to be taken with fewer unintended consequences. For example, the technological fix of using NETs, which is proposed as a cost-effective way to combat climate change, will have considerable social impacts that are foreseeable. For example, some of their applications will imply changes in land rights, displacement of people, and competition for food supplies (Beck and Mahony 2018). In this case, it means that those who have contributed the least to climate emissions will be severely affected not only by climate change, but also by measures taken to combat it. Thus, there is no way to bypass the public and political spheres when presenting technological solutions, because they need to be openly discussed. There is always a risk that unintended and negative consequences of proposals will be concealed or ignored, even though it is possible to gain knowledge of them.

Thus, when faced with complex issues and multiple proposals for how best to solve them, it is necessary to explore whose interests and what storylines and frames are reflected in the proposals, and to bring their wider consequences to the table. The next section will focus on science and scientific expertise, and their role in shaping our understanding of environmental issues. This is important not least because science is a driving force and navigational tool in environmental decision-making and regulation.

Claims and counterclaims in environmental storylines and politics

In politics, media reports, and public debates, we find very different evaluations of the current state of society and its development. There are stories of social and material progress and a belief that environmental problems will be solved without any deeper structural changes. And there are stories about environmental decay and growing social inequalities between and within countries. These stories are not developed and disseminated only by non-scientific actors. Science is also active in their creation and dissemination. Scientific storytelling refers to the creation and sharing of science-based stories that are told by scientists and are intended to influence a broad audience and guide their actions (Dahlstrom and Scheufele 2018, Kosara and Mackinlay 2013, Lidskog et al. 2020). Scientific concepts and measurements are translated to be meaningful for a wider audience, and the stories are often presented in graphs and indexes that summarize complex and broad processes of change (Latour 1987, Stone 2012). These stories are frequently reported in the media, because scientists have epistemic authority; that is, because of their position, they are trusted to give an accurate account of our current situation and where we are heading (Collins and Evans 2007, Lidskog and Sundqvist 2018a).

Are we facing environmental destruction or environmental progress?

A dominant theme in environmental debates is the story of how the world is hurtling toward global ecological catastrophe. In the 1960s, in response to widespread confidence in economic growth, an environmental science-based storyline developed about how nature was threatened by industrial activities to such an extent that all of humanity was in danger. Carson (1962/2002) warned that the continued use of chemicals (especially pesticides) would not only lead to the death of birds, but in the long run would also threaten human life. Hardin (1968) and Ehrlich (1969) argued that population growth would lead to overexploitation of natural resources and food shortages. Commoner (1971) predicted that production technologies using new materials (plastics, heavy metals, artificial fertilizers, etc.) would lead to more, not less, environmental pollution. A well-known story today is that of the *Anthropocene*. Originally designating a proposed new geological epoch in which human activities are geologically traceable, it has developed into a grand narrative of how human impacts now threaten fundamental life processes on Earth (Hamilton 2017, Lidskog and Waterton 2016). This narrative has successfully spread outside the scientific community, not only through environmental movements and government agencies, but also through cultural institutions such as museums and galleries, which have hosted exhibitions and artistic performances on the theme (Lidskog and Waterton 2018). The narrative is dynamic and works in many settings; however, most of its meaning has stabilized around a collection of graphs labeled *The Great Acceleration*, in which twelve different indicators, including human population, gross domestic product, fertilizer consumption, and water use, show dramatic increases, but are linked to major adverse environmental impacts, such as a high concentration of carbon dioxide in the atmosphere, global extinction of species, and ocean acidification (Steffen et al. 2011, 2015a).

The story of the Anthropocene is something that provokes fear, as it asserts that humanity is facing its greatest challenge ever and that rapid and extensive societal changes are needed to halt this trend. Titles such as the above-mentioned *Learning to Die in the Anthropocene: Reflections on the End of a Civilization* (Scranton 2015) and *The Uninhabitable Earth: Life After Warming* (Wallace-Wells 2019) certainly have the potential to cause fear. However, the Anthropocene narrative also offers glimmers of hope, as there is still time to act. The challenge lies in balancing between emotional messages of fear and of hope, in order to tell a story that creates space and incentives for action.

Alongside this story, however, there are other stories that offer diametrically opposed views of the direction of current global development. One of the best-known

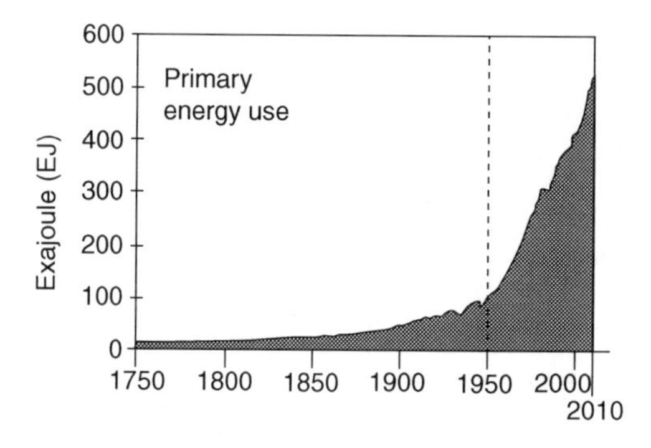

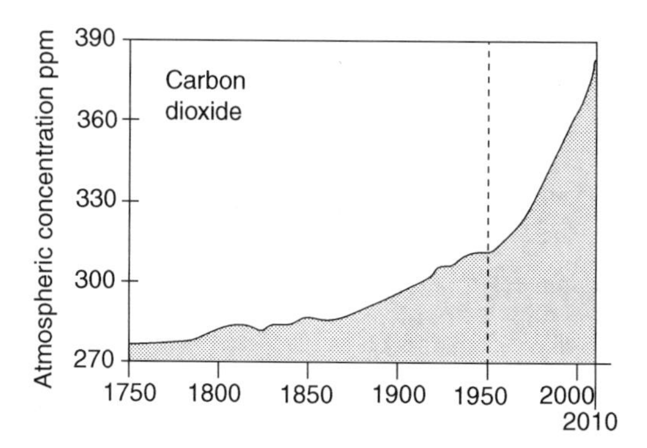

Figure 4.1 The Great Acceleration: Socioeconomic trends and global Earth system trends.

The illustration of Great Acceleration contains in total 12 graphs on increased human activities and 12 graphs on global environmental changes.

Source: adapted from Steffen et al. (2015a, pp. 84 and 87)

of these today is Steven Pinker's book *Enlightenment Now: The Case for Reason, Science, Humanism, and Progress*, which uses empirical data to demonstrate global progress in health, prosperity, safety, peace, and happiness (Pinker 2018). Similarly, the Gapminder Foundation, best known through Hans Rosling's TED talks (one of which has over 15 million views) and his book *Factfulness: Ten Reasons We're Wrong About the World – And Why Things are Better Than You Think* (Rosling et al. 2018), aims to fight devastating misconceptions about global development by producing free teaching resources based on economic and demographic statistics (gapminder.org). The essential storyline is that the world has become a better place to live in. Rosling shows that in a number of areas – such as health, education, and welfare – global trends give a very clear picture of increasing health and welfare. The problem, according to Pinker, Gapminder, and others who share this narrative, is that the vast majority of people have a distorted understanding of the world, believing that it is poorer, less healthy, and more dangerous than it truly is. By acquiring a fact-based worldview, we can "see that the world is not as bad as

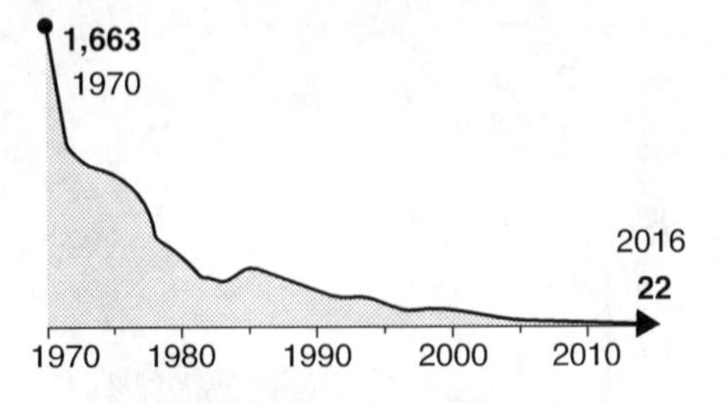

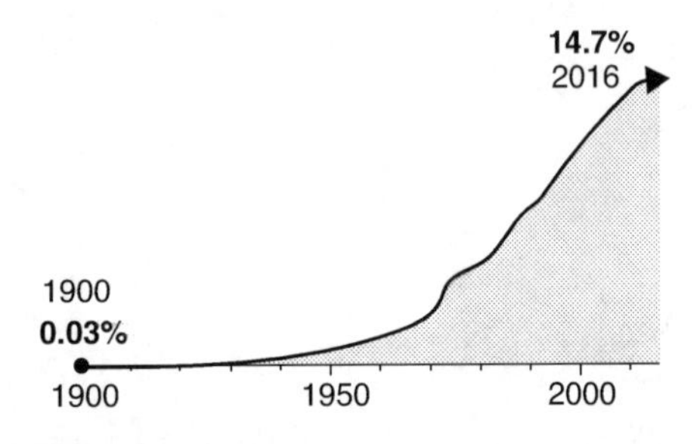

Figure 4.2 Examples of the improvements – bad things decreasing and good things increasing.

Source: adapted from Rosling et al. (2018, pp. 60–63). All 32 graphs can be found at https://www.gapminder.org/factfulness-book/32-improvements/

it seems – and we can see what we have to do to keep it making it better" (Rosling et al. 2018, p. 255).

The stories of ecological catastrophe and social progress are two cases of scientific storytelling about the state of the world, and both are examples of a common type of narrative structure that creates a sense that things are changing (Stone 2012, pp. 157–165). They tell very different stories, guide listeners in diverging directions, and evoke different emotions. However, they also share similarities, the most obvious of which is their claim to epistemic authority and belief in scientific facts. Both storylines claim to be based on established science, and they share the mission of spreading a scientifically based understanding of the state of the world. They illustrate their messages with dramatic graphs that show the directions of global trends with great certainty, and both storylines claim that the graphs are based on authoritative sources. All the figures they present support their main messages, thereby creating a strong sense of certainty about the direction society is heading.

Our aim with this chapter is not to claim that one story is true and the other false, but instead to show that scientific facts can be orchestrated differently; there are many ways to make facts serve the greater purpose of a storyline. The contrasting images presented in these cases are a good illustration of this. The advocates of one narrative may not question the numbers and data on which the other narrative is based. Instead, they might point to a bias in the selection of indicators and data, which may include facts and figures that support the wider narrative of society's current trajectory, and omit data that does not fit that narrative. They also might ascribe different meanings to the same data because they place different amounts of emphasis on social progress and environmental destruction. For example, the global increase in cell phone use is seen as a good thing in one narrative (Rosling et al. 2018, p. 62), but is associated with environmental destruction in the other (Steffen et al. 2011, p. 851). Thus, framing is important not only for the sense-making of people and organizations, but also for that of science.

This should not be interpreted as a form of far-reaching relativism, claiming that stories are all there is, and that there is no difference between scientific storytelling and other actors' storytelling. If that were the case, the story told by climate deniers would have to be considered just as valid as the one told by the IPCC and climate scientists. However, it should also not be taken the other way around: as meaning that scientific storytelling is always true because it is done by scientists. Again, we must stress that science alone is insufficient, and it is sometimes even unreliable. Also, we claim that the various facets of the social always matter, even when discussing scientific views on the current environmental situation and where society is headed. This is clear not least in the case of scientific expertise, that is, when discussing how science not only develops knowledge, but also provides advice to decision-makers.

Who is an environmental expert?

An expert is commonly defined as "someone who masters skills with recognized (indeed certified) competence which is called upon in decision-making processes" (Callon et al. 2009, p. 228). As this definition indicates, expertise has to do with *competence* – the mastery of specialized knowledge, skills, or practices, acquired through extensive training (Collins and Evans 2007). Possessing specialized competence is a necessary but not sufficient condition for being an expert. Expertise also has to do with *recognition* – being accorded expert status both by the expert community and by people and groups outside that community. Thus, expertise is constituted by both epistemic and relational factors; it concerns both specialized knowledge and social status (Lidskog and Sundqvist 2018b). Taken together, these two factors shape the social status of expertise, which gives a person, group, or profession epistemic authority within a particular knowledge domain.

Competence and recognition are related in a dynamic and complex way. Sometimes an expert in one knowledge domain is given expert status in other areas and begins providing expert advice on issues where he or she has limited or poor knowledge. Within environmental science, social scientists have pointed out that many natural scientists, who are highly qualified to comment on how nature works, are also asked about how society should change, or they take the initiative to make suggestions about how social practices and social institutions can be altered to combat environmental problems (Barry and Born 2013, Lidskog et al. 2022). A meteorologist may recommend new planning instruments, and an atmospheric chemist may speak in favor of green tax reforms, all in the name of halting environmental destruction. Of course, like all citizens, experts have the right to voice their concerns and develop proposals. But there is a risk that experts will use their expert status to legitimize positions that are far removed from their own areas of competence, or that they will neglect the complexity of problems and view them from a one-sided perspective. In the short term, this can lead to poor proposals that nevertheless influence deliberations and decisions. In the long run, it could threaten the epistemic authority of science. If experts push opinions on matters that are far outside their area of competence, or from an overly narrow viewpoint, it may fuel distrust in the epistemic authority of science.

Although the definition of environmental challenges has broadened to include not only environmental changes in a narrow sense, but also questions about their social drivers and possible solutions, the field of environmental experts is still dominated by natural scientists. There is a large flow of advice from scientists, who not only make statements about the condition of the environment, current emissions trends, and environmental consequences, but also propose measures for changing society. That can mean that measures are proposed without any informed social analysis, and without fully grasping the complex task of changing and transforming institutions. Often these measures are based on a technological fix. In other words, they are developed without considering the extent to which they are feasible, what their consequences would be, and who would be impacted by them. They lack a proper understanding of the root causes to why problems arise in the first place (Chapter 2), how they are distributed in society (Chapter 3), and what barriers they face (Chapter 5). When implemented, measures that seem relevant and effective on paper may be impossible to realize, or only possible to realize with severe unanticipated consequences. This can occur, for example, when a proposal fails to consider the fact that society is stratified – populated by people and organizations with different resources, situations, values, and interests.

Another problem arises if the tasks of changing social practices and transforming society are undertaken without considering how other fundamental goals – such as democracy and human rights – can be maintained and strengthened. An example of this is the concept of *planetary boundaries*, as it was originally formulated. This framework states that there are nine planetary boundaries, which give rise to nine thresholds for different biophysical subsystems and processes (Rockström et al. 2009). Within these boundaries there is a safe operating space for humanity. As long as the thresholds are not crossed, humanity has the freedom to pursue long-term social and economic development. This statement raises the question of whether other core values, such as democracy, human rights, and justice, can simply be disregarded with reference to planetary boundaries. In response to this, a proposal has been developed stating that the operating space for humanity should not only be safe, but also *just* (Hajer et al. 2015, Raworth 2012).[1] Raworth's famous doughnut model suggests that in addition to planetary boundaries, there are also social boundaries (concerning such things as social equity, gender equality, income and work, education, and housing) that define a socially just space in which humanity can thrive. This is an example of how proposals originally based on a physical science view of the environment need to be fertilized – complemented or radically revised – by contributions from the social sciences. Scientists involved in developing the planetary boundary framework have recently adopted this doughnut model and developed more systematic consideration of social issues. The Earth4All concept (Dixson-Declève et al. 2022) builds on the planetary boundary framework, but also emphasizes five "extraordinary turnarounds" that are needed to create well-being for all on a stable planet. These are eradicating poverty, counteracting escalating inequalities, strengthening women's participation and empowerment, producing healthy food in ways that do not harm ecosystems, and transitioning to clean energy. The authors say that this list of radical changes is not exhaustive, but it is interesting to note that they have recognized the need to seriously consider issues of social distribution/stratification.

Moreover, the need for social transformation is increasingly being stressed by various actors. It is recognized that social drivers and social activities have caused the current environmental problems, and that these activities must be changed. Therefore, international environmental expert organizations such as the Intergovernmental Panel on Climate Change (IPCC) and the Intergovernmental Science-Policy Platform on Biodiversity and Ecosystem Services (IPBES) are increasingly inviting social scientists to participate in their work, although natural scientists still dominate these organizations (Stenseke and Larigauderie 2018, Vadrot et al. 2018).

However, transformative change requires substantial input from the social sciences, because it is precisely society, with its institutions, regulations, and social practices, that needs to be transformed. The role of social science includes both contributing constructive, solution-oriented or, even better, transformation-oriented (see Chapter 6) knowledge, and ensuring that decisions and actions are not based on unsupported views of how society works (Beck et al. 2014, Jasanoff 2005, Jetzkowicz et al. 2018). Environmental problems involve socio-ecological dynamics, and therefore producing valid and relevant knowledge about environmental problems and how to solve them requires knowing about both nature and society, and not least about their interdependencies and interactions.

Does everyone have to become an expert?

We have stressed that environmental messaging is highly dependent on science. Scientific messages are communicated to the public in a variety of ways. The public is then expected to make sense of rather abstract problems using abstract terms such as "ppm" or "carbon dioxide equivalents," and they can use apps to calculate their "climate" or "ecological" footprint. Informing and educating the public is crucial, it is claimed, since everyone's actions play a part in the whole, and everyone has a share in the responsibility for preventing further destruction.

We have already stressed that emotions, values, and norms are important for people's sense-making about environmental problems. In addition, the already existing stock of public knowledge is crucial to consider. Existing knowledge is both a resource and a hindrance. The sociology of knowledge has found that we are all guided by knowledge and ideas that we internalize from the social context in which we live (Berger and Luckmann 1967). This is not just a matter of factual knowledge or knowledge that we learn through education. Much of our knowledge is about how things work on a practical level in everyday life. There is also a body of knowledge about how everything makes overall sense, which Berger and Luckmann call the "symbolic universe." This knowledge can include religious and spiritual beliefs, but can also include other sources of knowledge, such as scientific knowledge. Most of our knowledge consists of things we just take for granted; we rarely question them or make them objects of reflection. We see the world we live in as normal and natural, and we have ideas, concepts, and storylines that justify the order of things.

Audiences of environmental messages will interpret and make sense of these messages based on their pre-existing knowledge. Existing knowledge may change, but it usually cannot be completely replaced by something new. Hence, messages about environmental problems can gain acceptance if they connect in some way with people's understandings of the world. They may, however, also be rejected if they clash with the basic assumptions and views of existing bodies of knowledge. A problematic area in terms of the potential for environmental messages to gain traction with the public is the extremely high degree of differentiation and specialization in contemporary society, which is due to the long historical development of the division of labor and expertise. This far-reaching differentiation may make each of us a specialist within a very delimited area and a non-expert in many other areas, making it difficult for us to gain a comprehensive understanding of an issue or of society as a whole. In general, the public has learned to ignore the consequences of most of its actions. We do our part in the big "machine," but remain ignorant of the wider social and ecological consequences of our actions, regardless of whether they have to do with work or consumption.

However, this broader view is precisely what much ecological messaging invites us to try to understand. Many environmental storylines and framings try to construct a new "symbolic universe," a critical awareness of how "everything is interconnected." Concepts such as the planetary boundaries, the Anthropocene, sustainability, globalization, ecological footprints, climate justice, and so on have been invented to help us grasp something of this vast whole and how everything is connected. Visualization, through exhibitions, documentaries, and movies, can also help. However, creating such a comprehensive and holistic understanding of the problem is a huge challenge for everyone, given the complexity, specialization, and differentiation of society (see also Chapter 5 on barriers). We are all non-experts in most situations.

The sociology of knowledge usually distinguishes between different levels of knowledge and awareness. We receive huge amounts of information every day through mass media and social media, some of which we may process and some of which we may quickly forget (see Box 4.4). However, much of our knowledge is of a practical nature that we rarely reflect on. We have tacit knowledge about how to do things. This is embedded in the various social practices of everyday life (how to cook, drive, wash, use social media, etc.). Today, much of this knowledge is closely linked to expert systems that we are taught to use but not to understand. All kinds of technical devices help us get through the day. It is mostly when there is a disruption in service that we reflect upon or even question our practical knowledge and the tools that assist us. But again, such reflection is more the exception than the rule. Part of the environmental messaging concerns the importance of not just learning and accepting scientific facts and utilizing existing expert systems, but also of acquiring new practical knowledge. This includes sorting waste, adopting new consumption practices, commuting by bicycle instead of car, acquiring new (or rather reinventing old) skills and competencies such as repairing things and growing one's own vegetables, and even such seemingly simple things as borrowing and sharing things with each other (activities that are now digitally facilitated). We can see the outlines of a do-it-yourself (DIY) boom in parts of the population, but it is important to keep in mind that our current practical knowledge has very little to do with being literally self-sufficient. In fact, the difficulty or impossibility of a true DIY lifestyle is a consequence of the extremely high level of differentiation and specialization in society.

Box 4.4 How media shapes the environmental message

Where do we get knowledge from? A key area to consider is the mass media, including social media. An essential task of environmental sociology is to consider how the media shapes the public agenda and the everyday sense-making of environmental messages. It is clear that mass media plays a huge role in our understanding of environmental problems and solutions. First, it plays a powerful role in that it shapes what is on the agenda. Why, for example, is air pollution not a top news story, even though WHO has found that it causes more than four million premature deaths each year (WHO 2022)? Secondly, it shapes the messages, the storylines, and framings. Over the years, TV has played an extraordinarily important role in visualizing and dramatizing the environmental message (Hannigan 2022). However, due to its format (rapid production, limited space), dependence on newsworthiness, tendency to present polarized views, and other contextual factors, the messages are often framed in very specific ways. Although some reporting in news features and environmental documentaries can achieve a degree of depth, the general problem of simplistic, superficial, and biased reporting remains. This is accentuated if we consider how social media shapes the representation of environmental problems. Now anyone can be both a content creator and a mediator, spreading content generated by other social media users. What information should we trust? The importance of including social media in an analysis of popular understandings of environmental problems cannot be overstated, because people, not least in the younger generations, spend hours every day on various social media sites (Hruska and Maresova 2020). People receive messages from peers, influencers, and celebrities. They engage in conversations, seek to belong to communities, compare themselves to others, and seek social validation. The extensive use of social media stimulates ecologically unsustainable overconsumption, for example through messages from influencers and the tendency of social media users to display status consumption (Boström 2023). However, through the creation of online communities, social media can also be a source of inspiration for the development of ecological awareness and more sustainable everyday habits. In any case, the environmental sociologist needs to ask: how does this media format shape the understanding of environmental problems and solutions?

The learning challenges around sustainability issues are enormous (Boström et al. 2018). We are asked to become more like experts in everyday life, even though we are far from being so in most areas. In the final chapter (Chapter 6) we will return to this question of lifestyle transformation and consider the potential for transformative learning – a concept that brings together different aspects of norms, values, emotions, and knowledge.

Post-truth society, knowledge resistance, and climate denialism

The term post-truth has rapidly become part of public, media, and scientific debates (Sismondo 2017, van der Linden and Löfstedt 2019). In 2016, *Oxford Dictionaries* selected "post-truth" as its word of the year, defining it as "relating to or denoting circumstances in which objective facts are less influential in shaping public opinion than appeals to emotion and personal belief" (Oxford Dictionaries 2016). Today, some argue, post-truth is a central feature of contemporary society and social life (Iyengar and Massey 2019, Lewandowsky et al. 2017, Shelton 2020).

However, denialism is anything but a new phenomenon. There are numerous cases where organized interests have made massive efforts to counter warnings that a product or substance poses a threat to the environment and/or human health. In her classic book *Silent Spring*, Rachel Carson (1962/2002) claimed that the use of synthetic pesticides (not least DDT) would lead to a silent spring, a bleak world where birds no longer sing. She told a story that linked the use of chemicals to environmental destruction and threats to human health and life. Chemical companies immediately released a counter-story: that banning pesticides would lead to famine and human disease (Krupke et al. 2007).

This is not an isolated case. In their book *Merchants of Doubt: How a Handful of Scientists Obscured the Truth on Issues from Tobacco Smoke to Global Warming*, Naomi Oreskes and Erik Conway explore fifty years of what are now called post-truth strategies and organized denialism (Oreskes and Conway 2010).

Exploring a number of cases, they show how a relatively small group of scientists, with close ties to industry and politics, worked systematically to mislead the public. The strategy was to stress scientific uncertainty, keep scientific controversies alive, promote doubt, discredit science, attack scientists, and spread misinformation. They challenged studies that linked smoking to lung cancer, coal burning to acid rain, CFCs to ozone layer depletion, and carbon dioxide emissions to global warming. They insisted that the media must provide balanced reporting and give their views the same space and status as the established scientific ones.

The best known and most discussed case of this today is climate denialism. As a number of studies have shown, this amounts to a "denial machine" that, in a well-organized and systematic way (Dunlap and Brulle 2020), spreads its message of strong and far-reaching scientific disagreements and controversies about climate change. These are supposedly being silenced, either by scientists who are protecting their own (biased) research or by a global conspiracy that favors the discourse on climate change (Dunlap and McCright 2015). This message sows doubt about established (climate) science by presenting its own facts, figures, and arguments claiming that there is no such thing as human-induced climate change. They disseminate a storyline that there is no reason to take action on climate change, and that the measures proposed by the international community and national governments are costly, unnecessary, and will have large-scale negative impacts on society (harming human welfare and limiting individual freedom). This organized climate denial leads to the creation of a landscape populated by diverging and competing storylines, all of which claim to be based on science.

The climate denial movement originated in the USA but has spread internationally. Not least because of social media, climate denial messages are easily accessible everywhere. Some social media users find themselves in echo chambers where they only encounter messages and perspectives that reinforce a particular framing of the issue, which facilitates the formation of like-minded groups (Cinelli et al. 2021). The use of feed algorithms by social media platforms further reinforces these processes. Online users also tend to prefer information that supports their worldview and are more likely to ignore dissenting information, which leads to polarization. In their study of the US climate-denial movement, Dunlap and McCright (2015) find that it has created a parallel scientific universe in which scientific counterclaims are widely distributed through social media and social networks.

Four things are important to note here. First, climate deniers are extremely critical of established science but entirely uncritical when it comes to their own truth claims. Whereas the IPCC has a structured way of assessing and synthesizing scientific knowledge, including procedures for handling and communicating uncertainties, the climate denial movement claims to be based on (better) science, but does not foster any critical or reflective appraisal of its own assumptions, standpoints, and arguments. It is reflexive in its development of strategies for how best to spread climate skepticism and denialism, and strives to create blind trust among its followers about what is fake and what is true (Boström et al. 2017). In this sense, even if it presents itself as scientific and as much more valid than established climate science, climate denialism is anti-scientific, as it fails to foster healthy skepticism of all knowledge claims, including its own. As Robert Merton noted as early as the 1940s, one of the norms forming the ethos of modern science is "organized skepticism," the activity whereby researchers constantly question and scrutinize knowledge claims, including their own (Merton 1942/1973). Science is characterized by critical examination of its own knowledge production and transparency about how it judges knowledge claims to be valid.

Secondly, supplying more and better-disseminated information about climate change may be an insufficient strategy for countering climate denial. As Mikael Klintman (2019) shows in his book *Knowledge Resistance: How We Avoid Insight from Others*, it is naive to believe that fact resistance can be cured by providing more facts about the phenomenon in question. Instead of being viewed as a symptom of a lack of knowledge, fact resistance should be understood as serving a function for a person or group. Questions to ask are: why do they believe this? What benefits ("adaptive value") do they get by believing as they do? There are reasons why a knowledge claim is adhered to, and these lie not only in the quality of the claim, but also in the social function it serves. There are social incentives for believing a particular storyline and adopting a particular viewpoint, and these can pertain to such aspects as identities, social relations, and group belonging (facets one and two in our framework of the social). From the perspective of the sociology of knowledge, it should be emphasized that there are also social reasons for accepting the view on climate change held by the IPCC and climate science. As we discussed in the case of expertise, both epistemic and social factors are involved, and the challenge is not to stare oneself blind at one or the other of these factors, but to keep both in mind when evaluating knowledge claims, storylines, and the trustworthiness of organizations that make claims and disseminate storylines.

Thirdly, knowledge resistance has not emerged haphazardly. It is organized and orchestrated by actors with enormous power resources (see Chapter 2 on the three dimensions of power). In national and international politics, scientific knowledge has become an important resource for motivating action (Beck et al. 2014). This has led organized interests to put more effort into getting people to believe or disbelieve particular knowledge claims and storylines. For example, the US climate denial movement enjoys the political, economic, and media support of powerful actors (Dunlap and Brulle 2020). At front stage there are individuals and campaigns, but backstage there are investors, PR consultants, and think tanks that develop strategies for tailoring misleading or false information to people's situations, worries, and prejudices, and thereby sowing doubt about both climate change and the importance of addressing it. In Europe, too, there is a rich organizational landscape of think tanks involved in downplaying the seriousness of environmental problems and contributing to the "manufacturing of ignorance" (Almiron et al. 2021). By persuading target groups or the public to be uncertain about climate change or even to deny it, such actors obstruct the efforts of policymakers to develop far-reaching climate and environmental policies.

Fourthly, it should be mentioned that there is another kind of climate denialism and post-truth, namely that of knowing about climate change but doing nothing about it, or only picking the low-hanging fruit. As emphasized in Chapter 2, individuals, groups, and organizations have other aspirations and goals than environmental ones, and many environmental problems are complex and abstract, and require determined and coordinated action to solve. This creates space for a more subtle form of climate denialism that is embedded in people's habits and everyday practices. It gives rise to claims that what any single individual, group, organization, or nation-state does is of no consequence, or that until there is a binding global agreement on climate action, there is no reason to take the first step by changing one's own behavior (see Chapter 6 for a discussion of agency and change). One also might only commit to superficial changes without any real transformative potential. These and other social barriers to transformative change are discussed in the next chapter. Examples of this last type of denialism have also been called "discourses of delay," that is, framing strategies that justify inaction or inadequate action (see Box 4.5).

Box 4.5 Discourses of delay

Lamb et al. (2020) argue that in addition to outright denial of human-caused climate change and attacks on climate scientists and the scientific consensus, there are a number of strategies that can be described as "climate delay" discourses. Even if they partially affirm that climate change is occurring, they are essentially part of the climate countermovement. Here are four examples:

1. **Shifting responsibility.** This strategy involves claiming that someone else should act first. For example, "whataboutist" reasoning argues that other countries or sectors contribute more to greenhouse gas emissions and therefore have a greater responsibility to take action.
2. **Pushing non-transformative solutions.** The argument here is that disruptive change is not necessary. This is our main focus in Chapter 5, and it also relates to the frame of the technological fix (and the other fixes). This category also includes cases where the action taken is more about talk than real change, and where only carrots and no sticks are proposed.
3. **Emphasizing the downside of climate policies.** The argument here is that climate measures will be too costly. The consequences for society of taking climate action will be worse than inaction. Appeals to employment, public welfare, quality of life, and social justice may be part of the storylines. Fossil fuels are seen as irreplaceable for poverty reduction. Such storylines may attract low-income groups, marginalized sectors of society, or people in developing countries (see Chapter 3).
4. **Surrendering to climate change.** This is a storyline of impossibilism or even doomism. Impossibilism is the belief that it is not possible for society to organize large socio-economic transformations, and that minimal interventions should be supported. Doomism is the belief that any action we take will be too little, too late. We cannot avoid a climate catastrophe, and the only possible response is to adapt, or, in religious versions, to place our fate in "God's hands."

Conclusion

This chapter has emphasized the social processes of environmental (non) sense-making, that is, how society and social groups understand environmental problems. We have stressed that there are many different understandings of the world, and thus also of the environmental challenge – its severity, its causes, ways to deal with it, and who should be responsible for dealing with it. To analyze how people and organizations understand the environmental challenge and why they understand it the way they do, all five facets must be considered.

Inner lives: This theme concerns how people and groups make sense of environmental messages and form opinions. Knowledge is important, but so are personal values, internalized norms, emotions, and identities, as well as the larger storylines and frames that guide the interpretation of facts. Knowledge is important as part of a wider story that directs our attention, shapes our priorities, and guides our actions. This theme helps to highlight the importance of people's values and worldviews, and how they can help or hinder them in taking environmental action and learning to lead more ecologically sustainable lives.

Social relations and interaction: It is not the facts as such that matter, but the social meanings we attach to them, which are derived from social belonging and interaction. This concerns both the role of social norms and everyday discussions of environmental issues. It follows that there is an ongoing, sometimes highly polarized, struggle between social groups over how to understand the current environmental challenge. Therefore this theme helps to draw attention to the social roots of denialism and knowledge resistance, as well as to social communities and movements that seek to instigate change by creating and advocating alternative storylines.

Socio-material entities and social practices: This theme is concerned with issues of sense-making in terms of the provision of information, instruments (such as apps for calculating your carbon emissions or ecological footprint), knowledge, and education. It highlights the embeddedness of social life, and emphasizes that social-material entities are an integral part of everyday life. Because everyday life and social practices are often taken for granted, it is necessary to find ways to discover and reveal their environmental impacts.

Social stratification: When we argue that environmental problems must be understood as caused by society, we do not mean that there are a few simple social recipes for solving them. Society is highly complex and differentiated, environmental impacts are socially distributed, and knowledge is highly specialized. Moreover, social drivers and human motivations look different in different places and settings. We view society and the environment from very different vantage points in the social landscape. Thus, when we develop our understanding, we do so from particular positions and contexts.

Institutions: Science plays a key role, but it is not enough, as the chapter has shown. Storylines and framings can be based on science, but also on other sources. Economically powerful interests, including think tanks, work hard to frame environmental messages in particular ways, and some of them have been successful in casting doubt on the seriousness of problems such as climate change. Another important task for environmental sociology is to look at how various institutions in society are inclined to adopt a particular framing of problems and solutions, such as the technological, cognitive, and structural fixes.

Taken together, environmental framings and storylines flourish in society. For example, there are ideas about the necessity of economic growth and degrowth, about the potential of green technology and its risks, and about the need to radically transform society and the impossibility of doing so. Such storylines influence how we understand environmental challenges. There are also social norms about what constitutes a good and successful life that make us more or less willing to adopt a particular storyline. When an environmental problem is viewed from an abstract and universal perspective, it can look very different from when it is viewed from a local, everyday perspective. Where and how we live, what our social relationships are, and who we are in terms of gender, class, social position, and cultural belonging, are crucial factors to consider. Such social aspects influence what we know and believe in, what skills we have and lack, and what emotions we experience when we are the recipients of dramatic environmental messages.

Navigating this situation is not easy. Listening to the science, as Greta Thunberg suggests, is good and straightforward advice, but it is not enough and it is a bit simplistic. Yes, we do need to listen to science, but not in an uncritical and unreflective way. Knowledge, even scientific knowledge, contains not only facts, but also interpretations of those facts, and these involve normative commitments and emotions. The interpretations also stem from the deeper assumptions and worldviews about human life, society, and progress that we all carry with us, and thus they can lead to diametrically opposed storylines, such as those of Will Steffen and Hans Rosling (see Figures 4.1 and 4.2) – one warning that we are overshooting planetary boundaries and one celebrating global progress. Storylines and frames are at work everywhere, and we need to be constantly mindful of this fact. There is a middle ground between the extremes of a radically relativistic post-truth perspective (that it is not possible to evaluate the validity of a knowledge claim) and simple scientism (that science alone can provide guidance and certainty). Storylines and frames play their roles in justifying the various "solutions" that different actors construct and believe will solve the environmental crisis. The next chapter will critically review such proposed solutions and consider whether many of them should be seen as part of the problem rather than part of the solution. The final chapter will then discuss solutions that we believe have more promising transformative potential.

Questions for reflection and discussion

In the Introduction, we find three different stories about our environmental situation. Which of these stories do you think is most common in society, and why? Which one is most common in your context, among the people you spend time with? What are the reasons why a story is accepted or rejected?

The chapter emphasizes that knowledge is not enough to trigger action. Knowledge needs to be linked to norms, values, and emotions. Can you think of examples from your own life of situations where you know what is the right thing to do, but you still act differently? What do you think it takes for you to act on what you know to be true?

The authors present three common frames for handling an environmental problem, technological, cognitive, and structural fixes, and argue that all of these fixes need to consider the social. Do you agree with this standpoint, or are there environmental problems that can be solved by one or more of these fixes without considering the social?

Who is an environmental expert, and who is not? The chapter argues that both competence and recognition are required to be an expert. But what kinds of competence and recognition are important for being an environmental expert? How do you evaluate whether a person claiming to be an expert is actually competent? How do you judge whether a documentary you watch on the internet about, say, plastic in the oceans, gives an accurate description of the problem?

Box 4.4 discusses the role of the media and asks how the media shape our understanding of environmental problems and solutions. In your experience, to what extent have you gained an understanding of today's environmental challenges through your media use? What are the benefits and dangers of using social media as a way to gain knowledge about a particular environmental problem? Going back to Chapter 2 and its presentation of how power functions, what insights can the three dimensions of power contribute to our understanding of environmental problems and how they are framed?

Note

1 In their 2015 update in *Science*, they explain their view on society, arguing that politics is needed because the planetary boundaries framework does not dictate how societies should develop, only that these boundaries must be respected (Steffen et al. 2015b).

References

Almiron, N., Rodrigo-Alsina, M. and Moreno, J.A. (2021) 'Manufacturing ignorance: Think tanks, climate change and the animal-based diet', *Environmental Politics*, 31(4), pp. 576–597. https://doi.org/10.1080/09644016.2021.1933842

Arnold, A. (2018) *Climate change and storytelling: Narratives and cultural meaning in environmental communication.* Basingstoke: Palgrave.

Barbalet, J. (2002) 'Introduction: Why emotions are crucial', *Sociological Review*, 50(S2), pp. 1–9. https://doi.org/10.1111/j.1467-954X.2002.tb03588.x

Barry, A. and Born, G. (eds.) (2013) *Interdisciplinarity: Reconfigurations of the social and natural sciences*. London: Routledge.

Beck, S. and Mahony, M. (2018) 'The IPCC and the new map of science and politics', *WIREs Climate Change*, 9(6), e547. https://doi.org/10.1002/wcc.547

Beck, S., Borie, M., Chilvers, J., Esguerra, A., Heubach, K., Hulme, M., Lidskog, R., Lövbrand, E., et al. (2014) 'Towards a reflexive turn in the governance of global environmental expertise. The cases of the IPCC and the IPBES', *GAIA – Ecological Perspectives for Science and Society*, 23(2), pp. 80–87. http://dx.doi.org/10.14512/gaia.23.2.4

Beck, U. (1992) *Risk society towards a new modernity*. London: Sage.

Berger, P.L. and Luckmann, T. (1967) *The social construction of reality: A treatise in the sociology of knowledge*. London: Penguin.

Blühdorn, I. (2007) 'Sustaining the unsustainable: Symbolic politics and the politics of simulation', *Environmental Politics*, 16(2), pp. 251–275. https://doi.org/10.1080/09644010701211759

Blühdorn, I. (2011) 'The politics of unsustainability: COP15, post-ecologism, and the ecological paradox', *Organization & Environment*, 24(1), pp. 34–53. https://doi.org/10.1177/1086026611402008

Boholm, Å. (2015) 'Visual images and risk messages. Commemorating Chernobyl', in Å. Boholm (ed.) *Anthropology and risk*. London: Routledge, pp. 134–150.

Boström, M. (2012) 'A missing pillar? Challenges in theorizing and practicing social sustainability', *Sustainability: Science, Practice, & Policy*, 8(1), pp. 3–14. https://doi.org/10.1080/15487733.2012.11908080

Boström, M. (2023) *The social life of unsustainable mass consumption*. Lanham: Lexington Books.

Boström, M., Andersson, E., Berg, M., Gustafsson, K., Gustavsson, E., Hysing, E., Lidskog, R., Löfmarck, E., et al. (2018) 'Conditions for transformative learning for sustainable development: a theoretical review and approach', *Sustainability*, 10(12), 4479. https://doi.org/10.3390/su10124479

Boström, M., Lidskog, R. and Uggla, Y. (2017) 'A reflexive look at reflexivity in environmental sociology', *Environmental Sociology*, 3(1), pp. 6–16. https://doi.org/10.1080/23251042.2016.1237336

Callon, M., Barthe, Y. and Lascoumes, P. (2009) *Acting in an uncertain world*. Cambridge, MA: MIT Press.

Campbell, E.H. (2021) 'Corporate power: The role of the global media in shaping what we know about the environment', in K.A. Gould and T.L. Lewis (eds.) *Twenty lessons in environmental sociology*. New York: Oxford University Press, pp. 86–108.

Carson, R. (2002) *Silent spring*. Fortieth anniversary edn. Boston: Mariner Books, Houghton Mifflin Harcourt [originally published 1962].

Cinelli, M., De Francisci Morales, G., Galeazzi, A., Quattrociocchi, W., and Starnini, M. (2021) 'The echo chamber effect on social media', *PNAS Proceedings of the National Academy of Sciences*, 118(9), e2023301118. https://doi.org/10.1073/pnas.2023301118

Collins, H.M. and Evans, R. (2007) *Rethinking expertise*. Chicago, IL: University of Chicago Press.

Commoner, B. (1971) *The closing circle: Nature, man and technology*. New York: Knopf.

Dahlstrom, M.F. and Scheufele, D.A. (2018) '(Escaping) the paradox of scientific storytelling', *PLoS Biology*, 16(10), e2006720. https://doi.org/10.1371/journal.pbio.2006720

Dixson-Declève, S., Gaffney, O., Ghosh, J., Randers, J., Rockström, J., and Espen Stoknes, P. (2022) *Earth for all. A survival guide for humanity*. Gabriola Island, BC: New Society Publishers.

Dunlap, R.E. and Brulle, R.J. (2020) 'Sources and amplifiers of climate change denial', in D. C. Holmes and L. M. Richardson (eds.) *Research handbook on communicating climate Change*. Cheltenham: Edward Elgar, pp. 49–61.

Dunlap, R.E. and McCright, A.M. (2015) 'Challenging climate change: The denial countermovement', in R. Dunlap and R.J. Brulle (eds.) *Climate change and society: Sociological perspectives*, Oxford: Oxford University Press, pp. 300–332.

EASAC [European Academies' Science Advisory Council] (2018) *Negative emission technologies: What role in meeting Paris Agreement targets? EASAC policy report 35.* Halle, Germany: German National Academy of Sciences Leopoldina.

Ehrlich, P.R. (1969) *The population bomb.* New York: Ballantine Books.

Engdahl, E. and Lidskog, R. (2014) 'Risk, communication and trust: Towards an emotional understanding of trust', *Public Understanding of Science*, 23(6), pp. 705–717. http://dx.doi.org/10.1177/0963662512460953

Finucane, M.L. (2013) 'The role of feelings in perceived risk' in S. Roeser, R. Hillerbrand, P. Sandin, and M. Peterson (eds.) *Essentials of risk theory*, New York, NY: Springer, pp. 57–74.

Gates, B. (2021) *How to avoid a climate disaster: The solutions we have and the breakthroughs we need.* New York: Alfred A. Knopf.

Goffman, E. (1986) *Frame analysis: An essay on the organization of experience.* Boston: Northeastern Univ. Press [originally published 1974].

Grundmann, R. (2001) *Transnational environmental policy: Reconstructing ozone*. London: Routledge.

Hajer, M., Nilsson, M., Raworth, K., Bakker, P., Berkhout, F., de Boer, Y., Rockström, J., Ludwig, K., and Kok, M. (2015) 'Beyond cockpit-ism: Four insights to enhance the transformative potential of the sustainable development goals', *Sustainability,* 7(2), pp. 1651–1660. https://doi.org/10.3390/su7021651

Hamilton, C. (2017) *Defiant earth: The fate of humans in the Anthropocene.* Cambridge: Polity Press.

Hannigan, J. (2020) 'Media and the environmental movement in a digital age' in K. Legun, J.C. Keller, M. Carolan, and M.M. Bell (eds.), *The Cambridge handbook of environmental sociology*: *Volume I.* Cambridge: Cambridge University Press, pp. 193–205.

Hannigan, J. (2022) *Environmental sociology*. 4th edn. London: Routledge.

Hansen, A. and Machin, D. (2013) 'Researching visual environment communication', *Environmental Communication*, 7(2), pp. 151–168. https://doi.org/10.1080/17524032.2013.785441

Hansson, A., Anshelm, J., Fridahl, M., and Haikola, S. (2021) 'Boundary work and interpretations in the IPCC review process of the role of bioenergy with carbon capture and storage (BECCS) in limiting global warming to 1.5°C', *Frontiers in Climate*, 3, 643224. https://doi.org/10.3389/fclim.2021.643224

Hardin, G. (1968) 'The tragedy of the commons', *Science*, 162, pp. 1243–1248. http://www.jstor.org/stable/1724745

Heberlein, T.A. (2012) *Navigating environmental attitudes*. New York: Oxford University Press.

Hruska, J. and Maresova, P. (2020) 'Use of social media platforms among adults in the United States – Behavior on social media', *Societies*, 10(1), 27. https://doi.org/10.3390/soc10010027

IEA (2019) *Fossil fuel consumption subsidies bounced back strongly in 2018.* Paris: IEA. https://www.iea.org/commentaries/fossil-fuel-consumption-subsidies-bounced-back-strongly-in-2018

IUCN (1980) *World conservation strategy: Living resource conservation for sustainable development.* Gland: International Union for Conservation of Nature.

Iyengar, S. and Massey, D.S. (2019) 'Scientific communication in a post-truth society', *PNAS Proceedings of the National Academy of Sciences*, 116(16), pp. 7656–7666. https://doi.org/10.1073/pnas.1805868115

Jabbour, J. and Flachsland, C. (2017) '40 years of global environmental assessments: A retrospective analysis', *Environmental Science and Policy*, 77, pp. 193–202. https://doi.org/10.1016/j.envsci.2017.05.001

Jasanoff, S. (2005) *Designs on nature: Science and democracy in Europe and the United States*. Princeton, NJ: Princeton University Press.

Jetzkowitz, J., van Koppen, C.S.A., Lidskog, R., Ott, K., Voget-Kleschin, L. and Wong, C.M.L. (2018) 'The significance of meaning. Why IPBES needs the social sciences and humanities', *Innovation: The European Journal of Social Science Research*, 31(S1), pp. 38–60. https://doi.org/10.1080/13511610.2017.1348933

Klintman, M. (2019) *Knowledge resistance: How we avoid insight from others*. Manchester: Manchester University Press.

Kosara, R. and Mackinlay, J. (2013) 'Storytelling: The next step for visualization', *Computer*, 46(5), pp. 44–50. https://doi.org/10.1109/MC.2013.36

Krupke, C.H., Prasad, R.P. and Anelli, C.M. (2007) 'Professional entomology and the 44 noisy years since "*Silent Spring*". Part 2: Response to "*Silent Spring*"', *American Entomologist*, 53(1), pp. 16–26. https://doi.org/10.1093/ae/53.1.16

Lamb, W.F., Mattioli, G., Levi, S., Robbins, J.T, Capstick, S., Creutzig, F., Minx, J.C., Müller-Hansen, F., et al. (2020) 'Discourses of climate delay', *Global Sustainability*, 3, E17. https://doi.org/ 10.1017/sus.2020.13

Latour, B. (1987) *Science in action: How to follow scientists and engineers through society.* Cambridge, MA: Harvard University Press.

Lester, L. (2010) *Media and environment: Conflict, politics and the news.* Cambridge: Polity.

Lewandowsky, S., Ecker, U.K.H. and Cook, J. (2017) 'Beyond misinformation: Understanding and coping with the 'Post-Truth' era', *Journal of Applied Research in Memory and Cognition*, 6(4), pp. 353–369. https://doi.org/10.1016/j.jarmac.2017.07.008

Lidskog, R. and Sundqvist, G. (2015) 'When does science matter? International relations meets science and technology studies', *Global Environmental Politics*, 15(1), pp. 1–20. https://doi.org/10.1162/GLEP_a_00269

Lidskog, R. and Sundqvist, G. (2018a) 'Environmental expertise as group belonging: Environmental sociology meets science and technology studies', *Nature and Culture*, 13(3), pp. 309–331. https://doi.org/10.3167/nc.2018.130301

Lidskog, R. and Sundqvist, G. (2018b) 'Environmental expertise' in M. Boström and D. Davidson (eds.) *Environment and society: Concepts and challenges*. Basingstoke: Palgrave, pp. 167–186.

Lidskog, R. and Waterton, C. (2016) 'Anthropocene: A cautious welcome from environmental sociology?', *Environmental Sociology*, 2(4), pp. 395–406. http://dx.doi.org/10.1080/23251042.2016.1210841

Lidskog, R. and Waterton, C. (2018) 'The Anthropocene: A narrative in the making' in M. Boström and D. Davidson (eds.) *Environment and society: Concepts and challenges.* Basingstoke: Palgrave, pp. 25–46.

Lidskog, R., Berg, M., Gustafsson, K., and Löfmarck, E. (2020) 'Cold science meets hot weather: Environmental threats, emotional messages and scientific storytelling', *Media and Communication*, 8(1), pp. 118–128. http://dx.doi.org/10.17645/mac.v8i1.2432

Lidskog, R., Standring, A. and White, J.M. (2022) 'Environmental expertise for social transformation: Roles and responsibilities for social science', *Environmental Sociology*, 8(3), pp. 255–266. https://doi.org/10.1080/23251042.2022.2048237

Linnér, B.-O. and Wibeck, V. (2019) *Sustainability transformations: Agents and drivers across society.* Cambridge: Cambridge University Press.

Litfin, K.T. (1994) *Ozone discourses: Science and politics in global environmental cooperation.* New York: Columbia Univ. Press.

Merton, R.K. (1973) 'The normative structure of science' in R.K. Merton (ed.) *The sociology of science: Theoretical and empirical investigations.* Chicago: University of Chicago Press [originally published 1942], pp. 267–278.

Ojala, M. (2016) 'Facing anxiety in climate change education: From therapeutic practice to hopeful transgressive learning', *Canadian Journal of Environmental Education*, 21, pp. 41–56.

Oreskes, N. and Conway, E.M. (2010) *Merchants of doubt: How a handful of scientists obscured the truth on issues from tobacco smoke to global warming.* New York, NY: Bloomsbury Press.

Oxford Dictionaries (2016) Word of the Year 2016. https://languages.oup.com/word-of-the-year/2016/ (accessed 12 March 2021).

Pinker, S. (2018) *Enlightenment now: The case for reason, science, humanism, and progress.* New York: Penguin.

Raworth, K. (2012) *A safe and just space for humanity. Can we live within the doughnut?* Oxfam Discussion Paper, February 2012. Oxford: Oxfam International. www.oxfam.org

Ricœur, P. (1995) *Figuring the sacred: Religion, narrative, and imagination.* Minneapolis, MN: Fortress Press.

Rockström, J., Steffen, W., Noone, K., Persson, Å., Chapin III, F.S., Lambin, E.F. et al. (2009) 'A safe operating space for humanity', *Nature*, 461, pp. 472–475. https://doi.org/10.1038/461472a

Rosenbloom, D. (2018) 'Framing low-carbon pathways: A discursive analysis of contending storylines surrounding the phase-out of coal-fired power in Ontario', *Environmental Innovation and Societal Transitions*, 27, pp. 129–145. https://doi.org/10.1016/j.eist.2017.11.003

Rosling, H., Rosling, O. and Rosling-Rönnlund, A. (2018) *Factfulness: Ten reasons we're wrong about the world – and why things are better than you think.* London: Sceptre.

Schön, D.A. and Rein, M. (1994) *Frame reflection: Toward the resolution of intractable policy controversies.* Basic Books, New York, NY.

Scranton, R. (2015) *Learning to die in the Anthropocene: Reflections on the end of a civilization.* San Francisco, CA: City Lights Books.

Shelton, T. (2020) 'A post-truth pandemic?', *Big Data & Society*, 7(2). https://doi.org/10.1177/2053951720965612

Shove, E. (2003) *Comfort, cleanliness and convenience: The social organization of normality.* Oxford: Berg.

Sismondo S. (2017) 'Post-truth?', *Social Studies of Science*, 47(1), pp. 3–6. https://doi.org/10.1177/0306312717692076

Steffen W., Grinevald, J., Crutzen, P., and McNeill, J. (2011) 'The Anthropocene: conceptual and historical perspectives' *Philosophical Transactions of the Royal Society A*, 369(1938), pp. 842–867. https://doi.org/10.1098/rsta.2010.0327

Steffen, W., Broadgate, W., Deutsch, L., Gaffney, O., and Ludwig, C., (2015a) 'The trajectory of the Anthropocene: The great acceleration' *The Anthropocene Review*, 2(1), pp. 81–98. https://doi.org/10.1177/2053019614564785

Steffen, W., Richardson, K., Rockström, J., Cornell, S.E., Fetzer, I., Bennett, E.M., Biggs, R., et al. (2015b) 'Planetary boundaries: Guiding human development on a changing planet', *Science*, 347(6223), 1259855. https://doi.org/10.1126/science.1259855

Stenseke, M. and Larigauderie, A. (2018) 'The role, importance and challenges of social sciences and humanities in the work of the intergovernmental science-policy platform on biodiversity and ecosystem services (IPBES)', *Innovation*, 31(S1), pp. 10–14. https://doi.org/10.1080/13511610.2017.1398076

Stone, D. (2012) *Policy paradox: The art of political decision making*. 3rd edn. New York, NY: W.W. Norton & Co.

Thaler, R.H. and Sunstein, C.R. (2021) *Nudge: The final edition.* Penguin USA.

Thunberg, G. (2019) *No one is too small to make a difference*. London: Penguin Books.

UN (1987) *Our common future: World commission on environment and development.* Oxford: Oxford University Press.

Vadrot, A.B.M., Akhtar-Schuster, M. and Watson, R.T. (2018) 'The social sciences and the humanities in the intergovernmental science-policy platform on biodiversity and ecosystem services (IPBES)', *Innovation: The European Journal of Social Science Research,* 31(S1), pp. 1–9. https://doi.org/10.1080/13511610.2018.1424622

van der Linden, S. and Löfstedt, R. (eds.) (2019) *Risk and uncertainty in a post-truth society*. London: Routledge.

Wallace-Wells, D. (2019) *The uninhabitable earth: A story of the future*. Penguin Books.

Weick, K.E. (1995) *Sensemaking in organizations: Foundations for organizational science*. London: Sage.

Weick, K.E. and Sutcliffe, K.M. (2015) *Managing the unexpected: Sustained performance in a complex world.* 3rd edn. Hoboken, New Jersey: Wiley.

Weick, K.E., Sutcliffe, K.M. and Obstfeld, D. (2009) 'Organizing and the process of sensemaking' in K.E. Weick (ed.) *Making sense of the organization: The impermanent organization. Volume II.* Chichester: Wiley, pp. 129–152.

WHO (2022) *Ambient (outdoor) air pollution.* Fact sheet 19 December 2022. https://www.who.int/news-room/fact-sheets/detail/ambient-(outdoor)-air-quality-and-health

Wynne, B. (1992) 'Misunderstood misunderstanding: Social identities and public uptake of science', *Public Understanding of Science*, 1(3), pp. 281–304. https://doi.org/10.1088/0963-6625/1/3/004

Yearley, S. (1991) *The green case: A sociology of environmental issues, arguments, and politics.* London: HarperCollins.

5 Barriers

Social resistance through inadequate solutions

Ecolabels, green brands, carbon taxes, waste management, sharing economies, climate offsetting, bans on disposable plastics, circular economy, corporate social responsibility, eco-tourism, downsizing, transition towns, ad-free zones, car-free city centers, green growth, carbon capture technology, biodegradable plastics, and urban gardens. These are just a few examples from a long list of solutions to environmental problems, or to the socio-ecological crisis in general. Some proposals are abstract while others are concrete, and some are general solutions while others focus on specific environmental problems. The crucial question is: in what sense are they solutions? For whom are they solutions, and what do they aim to solve? Are they solutions to our socio-ecological crisis, or to our civilizational crisis? Or is it possible that this talk of solutions is deceptive and only serves to defend business as usual? Does it not substantially change harmful practices, but only slightly modify them? Should we talk about *solutions*, or is it better to talk about *transformations*?

We prefer to speak of transformation, because a transformed society is what we need. When different actors present their preferred solutions, they are often piecemeal, limited, inadequate, or even misleading. Rather than overcoming barriers to transformative change, they reflect barriers, and can even constitute barriers to transformative change.

This is not to say that there is no place for the concept of solutions, that is, solutions aimed at solving a particular problem. On the contrary, we can, must, and should seek solutions and speak in terms of both concrete and more theoretical solutions. Our position is that we should not approach problems with piecemeal solutions or engage in wishful thinking. What we should do instead is to develop a critical-constructive capacity to assess when packages of solutions are potential building blocks within a larger process of transformative change, and also to discern when they are merely talk, are not part of any deeper transformative change, or even might constitute barriers to transformative change. In this chapter, we will take a critical look at "solutions" and investigate how they can hinder rather than facilitate transformative change. That is why the theme of this chapter is "barriers." In the final chapter, we will present a more optimistic and constructive view on some solutions that have more transformative potential.

DOI: 10.4324/9781032628189-5

To transform society, we must focus not only on inventing, developing, and promoting activities that are sustainable, but also on finding ways to stop environmentally harmful activities. Otherwise we run the risk of having a lot of good environmental policies that work in one direction, but that are outweighed by policies that work in unsustainable directions. The problem, as Ingolfur Blühdorn (2011) puts it, is that current environmental policies are locked into "a politics of unsustainability," where there is much talk of change but little willingness to implement it. It is therefore very important to identify, examine, and find ways to overcome barriers to change. Unless barriers are identified and addressed, all the talk and ambition about transforming society will only be wishful thinking: strong words on paper but weak in practice. Contrary to the common belief that more and better-distributed environmental knowledge is the way forward, it is important to recognize that many unsustainable current practices are in fact supported by institutions and actors in society (see Chapter 2), often ones that claim to be committed to sustainability.

This chapter explores why it is difficult to initiate a more radical and far-reaching process of transformative change. It does so in three steps. In the first section, we highlight two common barriers to change: the complexity of society and resistance to change. These barriers make it difficult to implement solutions that are more than just partial and incremental. We use our model of the five facets of the social to address these barriers. In the second section, we discuss how the abundance of solutions risks becoming a further barrier to change rather than leading to transformative change. We discuss this in terms of piecemeal solutions and the "problem-solving" mindset. In the third section, we outline ten critical questions that we argue are useful to ask when something is presented by someone as a solution. We conclude by discussing the necessity of being critical of flawed solutions while remaining open to solutions with transformative potential, which is the subject of the concluding chapter.

Barriers to transformation

There is considerable uncertainty surrounding environmental problems, and especially about the proposed solutions. Even when there is agreement about the seriousness of climate change, species extinction, pollution, and other environmental problems, experts often disagree about how to respond. Environmental problems are often described as dilemmatic, ambiguous, or "wicked" (Head and Alford 2015, Höijer et al. 2006, Lidskog et al. 2018, Rittel and Webber 1973), which are terms used to describe problems that are complex, uncertain, and lack clear solutions. Existing knowledge about such problems is contradictory, there are conflicting goals, the stakes are high, and there are incompatible values and interests. There are no designated authorities charged with solving them. This "wickedness" is a major obstacle to transformation. Many actors ponder what to do in the face of the wickedness of environmental problems. Others may cite this wickedness to justify their inaction, and may even use it to obstruct the actions of others.

To better understand the barriers to transformation, we will discuss this wickedness by focusing on two interrelated social phenomena: the complexity of our (global) society, and resistance to change. The issue of power, and its flip side, powerlessness, is integral to both. In Chapter 2, we presented a range of theories on macro-, meso-, and micro-social levels that focus on the societal causes of environmental problems. We also emphasized that power is an integral part of the causes, because power explains why society continues to engage in environmentally harmful practices. Complementing this with a discussion of complexity and resistance to change allows us to include further insights about why it is difficult for societies to adequately deal with these root causes.

Paralyzed by complexity

What we want to emphasize here is that the perceived complexity and enormous scale of the environmental challenge leads to a kind of social paralysis that affects both individual and collective actors. It is very easy to become cognitively and emotionally incapacitated by the perception that one is just a tiny cog in the gigantic societal machine. But we must also stress that paralysis is not only a matter of perception, but is also a practical affair. Today's global interconnectedness makes it exceptionally difficult, in practical terms, to achieve societal change. The more complex a system is – in terms of the interconnections between its elements – the more difficult it is to change (unless it just explodes or implodes). Complex systems are difficult to oversee, and it is not easy to understand what should be changed, how it should be changed, to what extent the intended results will be achieved, and how to avoid significant unanticipated and undesirable consequences. Indeed, the multiple causes of environmental problems that we presented in Chapter 2 may seem overwhelming.

We can begin our investigation of complexity by looking at social stratification (facet four), which is related to institutions (facet five) such as the economy, international trade, politics, and science. Our complex global society involves societal differentiation on a global scale. For example, sociology has long been fascinated by the topic of the increasing division of labor, which has enabled economic and cultural development but also poses challenges to social integration. Modern society has evolved with an ever-increasing division of labor and specialization of expertise. This is what Adam Smith enthusiastically spoke of when he elaborated on what creates wealth in nations, and also famously suggested the metaphor of the invisible hand (Smith 1776/2008). The currently dominant neoliberal free trade doctrine has the same mindset. If everyone directs their aspirations and efforts in accordance with their particular skills and talents, and trade and markets are based on such a system of specialization, everyone in society will end up better off. This is also what sociologist Emile Durkheim spoke of a century later when he described how societies progress from mechanical to organic solidarity (from an undifferentiated society, where integration is based on shared values and beliefs, to a differentiated one, where the division of labor causes integration to be based on the need for each other's services), although this increasing differentiation makes it challenging to restore societal integration (Durkheim 1893/2013). In late-modern society, we inherit an extremely differentiated society in terms of labor, expertise, production, and consumption, on a global scale. Mainstream economics and neoliberal ideologies applaud the achievements that have resulted from this way of thinking. However, these developments have major social consequences in terms of social classes, status groups, and huge inequalities in life chances (see also Chapter 3). A high degree of societal differentiation and division of labor – within nations and worldwide – not only makes economic and social life unequal, but also makes it virtually impossible for any actor to have a broad overview of society. For example, due to globally stretched commodity chains, it is very difficult to get a proper understanding of the working conditions associated with most of the products that people consume in their everyday lives. Indeed, it is difficult to comprehend how many of the goods we take for granted depend on asymmetrical global power structures and unequal exchange on the global market, which Brand and Wissen (2021) call the "imperial mode of living" (see Chapter 2).

This division of labor and specialization of expertise has undergone rapid globalization in recent decades. International trade has a very long history, far predating the development of capitalism. Nevertheless, recent globalization processes and the capitalist world system (Wallerstein 2004) have implied a huge upscaling of intercontinental trade. Many companies outsourced their production to low-cost countries in the developing world decades ago. We now have a global system of opaque and highly complex commodity chains. This socio-material complexity (facet three) makes it extremely difficult for any actor, even a powerful multinational corporation, to keep track of what is going on behind the scenes in stretched commodity chains (Boström et al. 2015, Fridell 2019). Motivated by sustainability considerations, actors may try to manage the performance of suppliers along these commodity chains. They may refer to international institutions, rules and standards on sustainability and labor rights, codes of conduct, eco-certification, transparency programs, and the like. However, even if they are serious about improving sustainability (which may be doubtful, given that the outsourcing of production to developing countries with lax environmental standards and lower wages is mainly done to reduce production costs), they face several serious challenges (Locke 2013). Even a single item, such as a garment, is extremely difficult to monitor. This difficulty is inherent in the complexity itself, which includes the many actors involved in various stages of production and delivery of the item, as well as the physical, cultural, political, and linguistic distances between all of these actors. Many suppliers are themselves involved in complex networks of social interaction (Boström et al. 2015, Boström 2015, Börjeson and Boström 2018).

Social theories of various kinds note that interconnectedness between elements in society (actors/communities/technologies/politics/administration) creates locked-in mechanisms. It is difficult to change one element in a structure precisely because it depends on all the other elements.

Complexity makes it difficult to assess the effects of solutions at different scales. Solutions at one scale – such as technology to achieve resource efficiency or waste management – can be mistaken for solutions at a more general scale. Effective waste management does not solve the problem of global excess consumption. Energy and resource efficiency do not solve the problem of excessive use of energy or natural resources. The problem here is what is known as the "rebound effect"; gains from more efficient use of energy and natural resources per unit are outweighed by absolute growth.[1] The efficiency gain (unit savings) associated with an item leads to greater use of that item (direct rebound), and in addition, savings from efficiency gains imply that resources can be spent on other items (indirect rebound). For example, improved fuel efficiency in cars makes it possible to drive more – or to use the money saved from driving to consume more in other areas. Improved efficiency in the production of electricity may lead users to consume more electricity and the utility may invest in new facilities to increase electricity production. Thus, in the absence of regulation, efficiency gains can lead to much higher aggregate levels of energy use or resource outtake.

Contemporary society is also characterized by fragmented institutional structures (facet five), one example of which is increased political complexity. Democratically elected assemblies often make decisions that affect more than their own populations (Held 2004). In one sense they are too powerful, because they can make decisions that negatively affect people in other countries, such as allowing economic activities that cause pollution across borders. At the same time, however, they appear to be too weak, because they lack the power to deal with environmental problems that affect their own territories and populations. To address transboundary problems, a system of polycentric and multilevel governance has historically been institutionalized, either in the form of public-private partnerships or cooperation between different administrative units (municipalities, regions, states) (Lidskog and Elander 2010). When it comes to relevant actors for addressing environmental problems, the list can be long, including different actors at multiple political levels: the UN, the EU, central, regional, and local government authorities, economic actors, voluntary associations, and individual citizens. Complex and transboundary environmental problems have a multilevel, transnational, and multi-sectoral character that often requires a form of governance where the state or central government is not necessarily the only, or the most important, political actor (Brenner 2004, Speth and Haas 2006). As Beck (1992, 2000) notes, there are sub-political actors – outside the formal political system – that play an indispensable environmental watchdog role. Thus, it is unclear who has the capacity and responsibility to govern, not least when much of governance is conducted through networks.

There are often multiple understandings of what the problem is and how best to solve it (see Chapter 4). Scientific expertise is decentralized, fragmented, and challenged (Eyal 2019, Lidskog and Sundqvist 2018). Expert system and sociotechnical infrastructures are growing in different sectors such as energy, transportation, and food, which are interconnected and interdependent (Giddens 1990). Depending on how a problem is framed, different solutions are considered most effective and relevant. This is evident in the current debate on how to limit global warming. Should forests be used as a carbon sinks (by not cutting down trees), as building materials (replacing concrete), or as sources of biofuel (by maximizing growth and logging)? Is electrification of the transportation sector a way forward, or does it just shift environmental emissions from one sector (transportation) to another (energy)? Differences of opinion about environmentally efficient and politically viable remedies are not unique to policymakers. Such differences are also common within the scientific community.

In various facets of everyday life (inner life, social relations, social practices), people are surrounded by an overwhelming flow of information, including sensations, messages, and symbols. New information technologies add to the complexity of finding solutions by spreading a myriad of possible solutions to the public. How do we deal with the information overload? "More information" and appeals to our environmental conscience are simply not enough; on the contrary, they have become part of the problem. The multifaceted and flexible nature of language, rhetoric, images, and symbols adds to this complexity. What are we to make of the profusion of green symbols on product packages?

This information overload is accompanied by an overcluttered material landscape full of things and energy services that are integral to the functioning of our everyday social practices. Yet most of the social and environmental consequences of production and consumption are invisible to us (Chapter 2) or are mediated (Chapter 4). Take plastics and chemicals, for example. Plastic is everywhere. How can we do without it? How can we prevent plastic waste? How much bisphenol do I have in my bloodstream? And what is bisphenol anyway? Should I be worried? The ubiquity of plastic in our everyday lives and social practices makes it virtually impossible to live a plastic-free life, and this real and perceived impossibility is an obstacle to engaging in critical reflection on our contributions to plastic waste. The more we depend on something materially, the less we are able to see it, says anthropologist Daniel Miller (2010). Such things are so embedded in social practices, that they become part of what we take for granted. We can learn how to sort plastic waste, but not how to limit its inflow. Similarly, many people find themselves locked into the use of fossil energy. Knowing that it increases global warming is one thing, but knowing how to become less dependent on it is something else altogether. People and groups, as well as organizations and even states, are embedded in networks, socio-technical systems, and institutions that make it hard to find ways to change course and act more sustainably. Even if seemingly relevant paths are proposed, they may not be taken anyway, because there is no perceived or actual scope for individual action.

Complexity paralyzes inner lives (facet one). Complexity on a global scale means that finding solutions to environmental problems seems like an elusive and overwhelming task. For any actor, whether an individual consumer or a political representative, the situation of complexity risks creating paralysis. As we will show later, it can lead to a pragmatic but narrow mindset that addresses problems in a piecemeal fashion, rather than a mindset of transformative change. Dealing with the complexity of global society requires both a practical side (in terms of the economy, political structure, technology, social structure) and a visionary side – practical, because the increasing magnitude and scale of complexity makes change increasingly difficult to achieve, and visionary because extreme complexity makes change difficult to comprehend, both cognitively and emotionally. Complexity can seem overwhelming and induce a belief that one's actions are trivial and insignificant. It gives rise to feelings of powerlessness. Indeed, paralysis is an apt description of much of politics today. Gunderson (2022) uses the term "real helplessness" (which he contrasts with the psychological concept of "learned helplessness") to explain why people fail to take adequate climate mitigation actions in everyday life. The reasons are related to powerlessness, and to the lack of social organization, collective action, and adequate structures to confront the status quo.

A sense of one's smallness also makes it easy to shift responsibility to others. Because problems such as carbon emissions or plastic pollution are so complex and pervasive, everyone can blame someone else. This phenomenon is what Ulrich Beck (2009, p. 31) calls "organized irresponsibility," a situation where risky activities are regulated in such a way that, in the event of a disaster, no one can be legally held responsible due to the complexity of political hierarchies and organizational settings. The more actors that contribute to a problem, the less responsible they appear (see our discussion of "the problem of many hands" in Chapter 2). Indeed, even a large actor or powerful nation can point to its own relative smallness to argue that mitigation action is pointless. The problem is that even the largest industries and nations can perceive themselves as small in the face of massive problems, saying things like "the aviation industry is responsible for only 3% of global climate emissions," or "our nation is responsible for only a small fraction of total emissions." Hence, feeling overwhelmed by complex and massive problems easily leads to "impossibilism," which, according to Tim Jackson (2017, p. 221), is "the enemy of social change."

We certainly do not want to reproduce such impossibilism. Complexity makes change difficult, but not impossible. For example, despite the complexity of environmental problems, a number of problems have been solved. In some cases, there may be something akin to a "technological fix," a solution that does not entail any substantial changes in institutions or behavior (see Chapter 4). One example of this is the depletion of the stratospheric ozone layer, the solution to which was to replace certain substances (chlorofluorocarbons, CFCs) with less harmful ones (see Box 4.3, Chapter 4). Another is long-range transboundary air pollution; sulfur emissions have been dramatically reduced through technological developments, such as the introduction of low-sulfur fossil fuels and desulfurization equipment for coal-fired power plants (Lidskog and Sundqvist 2011).

These two areas were also characterized by complexity. They involved many actors with conflicting goals and considerable uncertainty, but in the end they were not overwhelmingly complex. Our point is that while solutions are rarely perfect, in the sense of completely eliminating an environmental problem, [2] there are cases of problems where it seems possible to deconstruct complexity, engage with trustworthy science, attribute causes and responsibility, develop alternative technologies, gain media attention for the problem by using resonant framings (such as the "ozone hole" or "forest dieback"), and mobilize public opinion, political support, and international cooperation.

In other areas, the knot of complexity is much harder to untie. These include the threat of mass extinction of species, plastic pollution in the oceans, chemical pollution in general, and climate change. In these areas, too, there are political institutions, international cooperation, international expert organizations, significant media attention, public campaigns by a wide variety of movements and organizations, trustworthy science, technological alternatives, and public opinion. Nevertheless, the problems are escalating. However, to say that these areas are more complex is not to say that they are impossible to address. The United Nations Agenda 2030 for sustainable development, with its 17 Sustainable Development Goals (SDGs), is a structured plan to deal with extreme complexity and massive sustainability challenges. It is the world's most comprehensive and universal policy agenda, has led to ambitious efforts, and has provided a general frame and platform for concerted and cross-sectoral action. Even if the goals cover different areas, they are shown to be part of the same problem. For example, climate change (SDG 13) is caused by production and consumption patterns (SDGs 7, 12) that affect harvests, freshwater, and biological life (SDGs 1, 2, 6, 14, 15), resulting in ill health, inequality, and uneven development (SDGs 3, 5, 10), which require concerted efforts by inclusive societies to solve (SDGs 16, 17)

Box 5.1 Agenda 2030 and the 17 Sustainable Developments Goals

The 2030 Agenda for Sustainable Development is a resolution adopted by all United Nations Member States in 2015. Its core is its 17 Sustainable Development Goals (SDGs), which are a global call to action to end poverty, protect the planet, and ensure that by 2030 all people enjoy peace and prosperity.

1. **No poverty:** End poverty in all its forms everywhere.
2. **Zero hunger:** End hunger, achieve food security and improved nutrition, and promote sustainable agriculture.
3. **Good health and well-being:** Ensure healthy lives and promote well-being for all at all ages.
4. **Quality education:** Ensure inclusive and equitable quality education and promote lifelong learning opportunities for all.
5. **Gender equality:** Achieve gender equality and empower all women and girls.
6. **Clean water and sanitation:** Ensure availability and sustainable management of water and sanitation for all.
7. **Affordable and clean energy:** Ensure access to affordable, reliable, sustainable, and modern energy for all.
8. **Decent work and economic growth:** Promote sustained, inclusive, and sustainable economic growth, full and productive employment, and decent work for all.
9. **Industry, innovation, and infrastructure:** Build resilient infrastructure, promote inclusive and sustainable industrialization, and foster innovation.
10. **Reduced inequality:** Reduce inequality within and among countries.
11. **Sustainable cities and communities:** Make cities and human settlements inclusive, safe, resilient, and sustainable.

12. **Responsible consumption and production:** Ensure sustainable consumption and production patterns.
13. **Climate action:** Take urgent action to combat climate change and its impacts.
14. **Life below water:** Conserve and sustainably use the oceans, seas, and marine resources for sustainable development.
15. **Life on land:** Protect, restore, and promote sustainable use of terrestrial ecosystems, sustainably manage forests, combat desertification, and halt and reverse land degradation and halt biodiversity loss.
16. **Peace, justice, and strong institutions:** Promote peaceful and inclusive societies for sustainable development, provide access to justice for all, and build effective, accountable, and inclusive institutions at all levels.
17. **Partnerships for the goals:** Strengthen the means of implementation and revitalize the Global Partnership for Sustainable Development.

The goals are extremely broad and are further specified through 169 targets and 231 global indicators. The goals are interlinked and must therefore be implemented in an integrated manner to tackle complex sustainability challenges. According to the UN, achieving the goals will require societal transformation at the national and global levels. Agenda 2030 emphasizes the need for global solidarity – that countries have a responsibility to help reduce environmental problems in other countries.

Each year, an annual *SDG Progress Report* is presented, which provides a global overview of the implementation of the Agenda 2030 using the latest available data. The UN Secretary-General presents an annual SDG Progress Report, which is developed in collaboration with the UN System, and is based on the global indicator framework and data produced by national statistical systems and information collected at the regional level. Recently and currently, the COVID-19 pandemic, ongoing conflicts in a number of areas, and increased climate change have resulted in cascading and interlinked global crises which have had strong negative effects on global development (UN 2022).

Source: UN (n.d.)

Resistance to change

We need to comprehend the complexity of global society to understand the barriers to transformative change. But the great complexity that exists on a global scale is not the whole picture. We also need to address resistance to change. Of course, perceived powerlessness due to complexity contributes to resistance to change. But there are other causes of resistance to change as well. Resistance to change is close to all of us. It is everywhere, in all spheres and domains of society. It is found not only among those who have the most to gain by keeping society as it is, especially the richest segment of the population (see Chapter 3), but also among other segments, because keeping things as they are is something that most people value. An analysis of resistance to change must therefore take into account both active and passive, or intentional and unintentional, resistance, as well as the role played by both actions of power (actors using resources to resist change) and actions of powerlessness. Complexity, perceptions of smallness, powerlessness, power, and resistance to change all interact. We will discuss this by linking the question to the five facets of the social.

Inner life. To begin with, we have many goals that run counter to combating climate change and other environmental problems (see Chapter 2 on microsocial causes). Such goals do not just change overnight based on new information. Moreover, regardless of whether you are privileged or marginalized, resistance to change appears in your inner life. The concept of "ontological security" (Giddens 1990, p. 92) refers to people's need to feel that their everyday lives are stable. You feel more secure if you sense that the things you know and value today will be the same tomorrow. We do also appreciate change in our lives, but not if it is too overwhelming, unpredictable, or beyond our control. We are socialized into systems of social norms and meanings as well as behaviors and practices that seem normal and natural to us. We form our hopes and expectations, such as what we can expect in terms of work, consumption, and social relations, in terms of the society in which we live. Sociological concepts such as habitus, practical consciousness, routinization, normalization/naturalization, and rituals indicate that much of our lives are unreflective – we do and feel things without thinking about why. Change is often threatening. People who experience helplessness or powerlessness, and thus feel that they cannot change things, may also tend to internalize existing conditions and consider them natural and even desirable.

Such internal resistance to change should not be interpreted as meaning that human beings are conservative by nature, always preferring the status quo and viewing change as something negative. People also strive for change and progress, to transform their surroundings, and make life better and more meaningful. Change can be exciting. We may seek adventure, innovation, and new experiences. Hence, both stability and change are important to people. However, not all changes are welcome, not even ones that are potentially good for an individual, a group, or society at large. An exciting new experience may be welcomed by someone whose everyday life is otherwise stable, but because of resistance to change, people may use sets of arguments – justifications – to defend their current way of life (see Box 5.2).

Box 5.2 Justifications for doing nothing? The case of defending overconsumption

An important traditional area of sociological inquiry has been to understand why and how populations defend and legitimize the existing institutional order of societies and communities, as well as their everyday practices (see, e.g., Berger and Luckmann 1967/1991, Boltanski and Thévenot 2006). In relation to this broad topic we can ask: Why do people in wealthy contexts continue to engage in ecologically unsustainable levels of consumption – overconsumption – despite increasing awareness of their climate and ecological footprints? When people feel unable to change, or if they resist change, they may use various arguments to defend and legitimize their overconsumption. Boström (2023) summarizes nine defensive arguments:

1. **The small agent argument.** Neglecting one's role with reference to one's smallness, and externalizing responsibility to larger institutions and infrastructures.
2. **The caring argument.** Defending overconsumption by referring to the need to care for others, which the person believes requires the purchase of commodities and services.
3. **The social comparison argument.** Blaming others, for example the super-rich or upper classes, who have significantly larger carbon/ecological footprints than one's own.
4. **The I-can-change-later argument.** Putting off changing one's lifestyle to become less dependent on consumerism until some distant future, when one believes that there will be more time to deal with the problem.
5. **The entitlement argument.** Referring to one's legitimate right to (over) consume, also as compensation for hard work and other sacrifices.
6. **The just-in-case argument.** Referring to the need to own and accumulate a buffer of goods that are never, or only infrequently used, but which may be considered necessary to have just in case.
7. **The minor harm argument.** Toning down the problem of one's own overconsumption, for example by actively avoiding relevant knowledge and information.
8. **The I-do-my-bit argument.** The belief that environmentally harmful consumption can be compensated and balanced out by something environmentally benign.
9. **The we're-doomed-anyway argument.** The fatalistic argument that we might as well continue as usual because the catastrophe is unavoidable anyhow.

Social relationships/interaction. Change can also be socially threatening. The norms, expectations, and consumption patterns of our existing society are embedded in our relationships. Because social relations are involved in reproducing environmental problems (Chapter 2), they also create resistance to change. An individual who learns that something is wrong with society and social life, and who feels a need to question some existing norms, values, assumptions, and habits, may have to question their own social affiliations and relationships. This can be very difficult, however, because humans are fundamentally social. One of the greatest threats to people's identity and well-being is the disruption of their social relationships, identities, and group belongings. The worst thing that can happen is to be misrecognized, excluded, labeled as deviant or extreme, and no longer be welcome in the social groups to which the person belongs (the "primary group"). Consider how hard it has been, and still is in many social settings, to become a vegetarian/vegan and stop eating meat. As studies of knowledge resistance have shown, people value their social belongings so highly that it is very difficult for them to question things that are taken for granted by group members (Klintman 2019). A complicating factor is that people orient themselves not only to existing norms, but also to what they *believe* to be existing norms (Cialdini 2007). If people believe that a particular norm, such as eating meat, is what most people, or the group to which they belong (or wish to belong) accept – even if no one has explicitly stated the norm – then they are likely to conform to that perceived norm.

Socio-material entities and social practices. The socio-material infrastructure and related social practices are among the complex factors that lock people into path dependencies. As was discussed above, the material landscape surrounding social life makes it difficult to change in a practical sense, as well as in a cognitive, visionary, and emotional sense. The existing socio-material infrastructure – especially its less visible aspects, such as water distribution, energy supply, and internet access – is taken for granted, making it difficult to imagine and try out new patterns of action. The city has its buildings and streets. We all live in our homes, unless we are homeless. Cities remain where they are, even though we can tear down buildings and build new ones and bike lanes. Homes can be renovated and made more energy efficient. Changing the built environment, even toward greater sustainability, requires an influx of new materials and an outflow of waste, which in turn requires energy. We may be willing to reduce our electricity and water consumption, but not to move to a smaller apartment, take shorter showers, spend less time on the internet, and stop streaming movies. Our resistance to change is built into our bodies, our habits, our relationships, and our buildings. We demand constant inflows of energy and materials, and we are unable to see how this demand increases over time (Rinkinen et al. 2021, Shove 2003). Consider how difficult it is to use less water. Are you ready to replace your toilet with an outhouse, and shower only once a week? Think about how difficult it is to get rid of plastics and environmentally harmful chemicals. Producers of textiles and outdoor equipment convince each of us that plastics and chemicals are necessary to make products more durable, flame retardant, and water repellent. Will you choose a less waterproof jacket because it contains fewer environmentally harmful chemicals?

The rigidity of social stratification. Why is greater equality so hard to achieve? Why, for example, do low-income citizens vote for leaders whose policies will not improve their situation? Questions like these are not easy to answer, but it is important to stress that social stratification itself creates resistance to change.

Of course, it is not difficult to see the role of vested interests with vast concentrations of power resources. Those at the top of the social ladder obviously have a lot to gain from keeping things as they are; although they should also be very concerned about global crises such as climate change, pandemics, and mass extinction of species. They also have reason to worry that rising inequalities may lead to a more socially unsafe society (Wilkinson and Picket 2011, 2018). Nevertheless, for decades, lobby groups have worked hard and with astonishingly unscrupulous methods to sow doubt about human-induced climate change, so that the fossil fuel industry can continue to operate profitably (Oreskes and Conway 2010, Kenner 2019).

The sociologically more difficult question is why do many groups lower on the social ladder also express climate denial, for example (Norgaard 2011). The problem with hierarchy is that for most people, there is always someone below you. Will change put you at risk of falling even further down the social ladder? People at the top have the most to gain, relatively speaking, from keeping structures as they are. But a considerable number of groups below them reason in similar ways, not least the middle classes in places like China, Vietnam, India, and Latin America, which have grown rapidly in recent decades (Brand and Wissen 2021, pp. 114–132, Hansen 2023). They too do not want to lose their (recently gained) privileges, material standards, and social position. In addition, more disadvantaged, and low-paid groups, who struggle to make ends meet, are also part of a consumer society, acquire gadgets and things, and compare themselves with others. They may feel that they have worked hard to achieve them and not want to risk losing them.

This perspective also includes marginalized groups, poor communities, and developing countries. In Chapter 3, we presented Robert Bullard's study *Dumping in Dixie: Race, Class and Environmental Quality* (Bullard 2000). This study found a strong relationship between the siting of hazardous waste landfills and the racialization and socioeconomic status of the surrounding communities. Industrial firms viewed black communities as the easiest places to site landfills, because of their perceived low levels of community organization and environmental awareness. Bullard finds several historical and socio-economic factors that explain why people in these communities had fewer opportunities to move and find jobs elsewhere, and why it was harder for them to oppose the landfills. He also shows that public policies and industrial practices support this shifting of costs from the rich to the poor. Studies in South Africa similarly show that the macroeconomic development path chosen by the post-apartheid government focused on large-scale industrial development to generate wealth. The vast array of social problems (including unemployment, poverty, and segregation) with which it struggled made environmental issues secondary (Leonard 2009, World Bank 2018). Many communities were primarily concerned with socio-economic issues, and there was little government or industry support for considering environmental impacts at the local level (Leonard and Lidskog 2021).

Thus, one reason why those who clearly stand to benefit from stricter environmental regulation do not support it is that they risk losing their jobs. They also have limited access to environmental discourses that problematize the activities that generate these jobs. Furthermore, the example of the yellow vest protests (see Box 3.3, Chapter 3) shows how difficult it is to achieve successful policy reform when policymakers fail to consider aspects of distributional justice.

Institutional inertia. Institutional theory points to various cognitive, regulatory, and material factors that serve to maintain the status quo. Institutional inertia is related to several of the issues mentioned above, such as how complexity and interconnectedness create lock-in effects. Resistance to change is also related to prescribed behavior. The reason we do not change our behavior is because powerful cognitive, normative, and coercive forces tell us not to. This is where complexity, power, and resistance to change interact. Complexity makes it overwhelmingly difficult even to comprehend change at the institutional level. Moreover, power structures and norms tell us not to question how issues are addressed and organized. Some institutionalized goals – such as growth – that run counter to saving the planet (see Chapter 2), are taken up by powerful elites and forcefully propagated in societies. The "politics of unsustainability" (Blüdorn 2011) is a major obstacle, with much policy development continuing along the trodden path, leading to only limited change.

Institutions also provide institutionalized means and resources to deal with problems in predefined ways. Problems with market functioning shall be solved by market means (taxes, subsidies, pricing mechanisms, etc.). Problems with capitalism shall be solved by capitalist means. Problems with regulations shall be solved with new regulations. State power is used to deal with problems of state power. Such institutionalized means and resources encourage a narrow problem-solving mindset, not a mindset of structural change. We are trained to resist structural change and to welcome a problem-solving mindset.

In summary, complexity and resistance to change involve all five facets of the social. While resistance to change may be most obvious and easiest to understand when it occurs among elite actors, the sociological analysis of resistance to change must include society as a whole. In Box 5.3, we highlight a journal article that offers a holistic explanation of why global society has failed to bend the global climate emissions curve despite several decades of climate mitigation efforts. In explaining this societal failure, the reader can see that all five facets of the social are involved.

Box 5.3 Why do global climate emissions continue to increase?

In a recent article published in the *Annual Review of Environment and Resources*, 23 climate social scientists ask why, despite three decades of climate mitigation efforts, global society has been unable to bend the global climate emissions curve (Stoddard et al. 2021). These experts identify nine overarching barriers that reflect the five facets of the social very well and are linked with both the complexity of global society and resistance to change. They assert that power permeates all of them.

The first three are metaphorically referred to as the *Davos cluster*, as they evoke the institutional power structures associated with the economic and political elite who meet each year in Davos to attend the World Economic Forum:

- International climate governance. Obstructionism and various failures related to international climate negotiations.

- Vested interests of the fossil fuel industry. This includes lobbying, organized denial, and climate delay strategies. The vested interests include both private and public actors.
- Geopolitics and militarism. Climate change is linked to historical and contemporary forms of military colonial/imperialist power.

The second cluster is called the *Enabler cluster*, and it acts as a legitimizing collaborator of the Davos cluster. This cluster frames the future as a simple extension of today, even though it could potentially be an important source of rapid change.

- Economics and financialization. Despite the existence of a range of economic theories, a neoclassical orthodoxy, which is increasingly allied to neoliberalism, predominates. Through a series of questionable assumptions, this orthodoxy manages to convince actors that the costs of mitigation exceed the costs associated with climate change.
- Mitigation modeling. Climate models are reliant on large-scale carbon dioxide removal, which does not yet exist, and which is part of a broader culture of technological optimism (see Technological Fix, Chapter 4).
- Energy supply systems. Renewable energy has so far only supplemented, not replaced, fossil fuels. Energy supply continues to be dominated by fossil sources.

The third cluster is called the *Ostrich and Phoenix cluster*. This is also oriented toward the status quo, although there is considerable potential for change and counterpower.

- Inequity. Increasing inequity (see also Chapter 3) decouples the vulnerable from the powerful, erodes the social trust required for collective action, and reinforces the preferences of elites (vested interests).
- High-carbon lifestyles. Due to psychological, social, contextual, and structural factors, people are locked into lifestyles and do not change them, even when they are concerned about the environmental consequences (see also Chapter 2).
- Social imaginaries. There is an epistemological monoculture. Visions of the future are dominated by a neoliberal frame and constrain the "social imaginaries." Deep transformative change is perceived as impossible. There is no cultural alternative (see also Chapter 4).

The third cluster has the potential to serve as a counterpower and trigger change as a domain of bottom-up experimentation, new norms, cultural change, and social movements. "It is conceivable that this cluster could begin to redefine the boundaries of analysis that inform the Enabler cluster, which in turn has the potential to erode the legitimacy of the Davos cluster" (p. 659). We discuss these issues in Chapter 6.

Piecemeal solutions and the problem-solving mindset

It is imperative that environmental scientists and other analysts maintain a critical and constructive attitude toward anything that is presented as a "solution." Various social actors propose solutions while having all kinds of reasons to avoid considering the root causes (Chapter 2) and structural problems underlying our social-ecological crisis. Rather than seeking real and deep transformation, however, many solutions legitimize business as usual, or only slightly modify existing operations, to be able to claim that the operations can continue. It is crucial to remember that activities presented as solutions may very well be part of the problem or serve to exacerbate another problem. It is also important to point out cases where certain solutions have the long-term potential to be woven into a promising transformation process. However, a warning is in order: the word "transformation" is gaining popularity and has become a buzzword (Lidskog and Sundqvist 2022). Like any concept, transformation can easily be misused or watered down in environmental discourse (see Box 5.5).

Of course, it is not easy to distinguish false solutions from genuine ones that can contribute to transformative change. While we cannot offer an easy-to-use manual or yardstick, we do develop some important principles and perspectives that can be used to evaluate proposed solutions, their importance, and their potential effects. It should be clear that many concrete, ready-made solutions such as eco-labeling, carbon offsetting schemes, and sharing economies can be invoked for both honest and deceitful reasons. We are not saying that some environmental solutions are false while others are true. The analysis needs to be more careful, nuanced, and cautious than that. It must go beyond superficial levels of blind applause on the one hand, or cynical hypercriticism on the other.

Sociology does not usually look at solutions in a literal sense. Sociologists open up seemingly fixed phenomena, to show that there are other ways of understanding them. This is also the case with proposed solutions. If someone says something is "environmentally friendly," it would not be surprising for an environmental sociologist to be a bit suspicious and critically examine what is meant by that claim. How can something be environmentally friendly? Perhaps it benefits the person making the claim rather than the environment. Or maybe it is just a strategy to sell a product or to make people feel more positive about a product, activity, or organization.

What is a solution, then? Basically, sociology defines a solution as anything that someone presents as a solution to any kind of a problem, large or small. Actors create and frame "problems" and they create and frame "solutions." Problems and solutions are thus "socially constructed." They are embedded in ideologies, policies, discourses, action repertoires, and so on. This is not to say that problems and solutions are nothing but words and thoughts, that they are fake, only exist in our minds, are not serious, and so forth. They are certainly real. If someone says she is hungry and suggests that the solution is food, the sociologist does not question the validity of that claim. The sociologist instead asks questions about the social circumstances behind the claim, in particular how it is framed. Is this problem – a person's hunger – framed as an individual problem, caused by the person's wrong prioritization, lack of knowledge, or lack of resources? Or is it framed as a social problem, caused by structures of inequality and uneven development at the local, national, or global level? As we stress in Chapter 4, how a problem is framed also determines what solutions are sought.

Solutions can activate different kinds of actors and institutions: the market, the state, civil society, and the actors involved in these fields (businesses, political parties, public agencies, research communities, social movements, and households). They can rely on voluntary compliance or mandatory regulation. Solutions can be connected to the local or the global level. They can be temporary (such as a temporary tax to phase out a substance) or long-term (such as a disposal site for nuclear waste). They can be concrete (a law on disposable plastics) or abstract (promoting a concept such as circular economy). We may speak about a particular solution to a particular problem. Or we may speak of standard solutions applied from case to case, such as eco-certification schemes or subsidies for energy saving in households. Solutions can also take the form of socio-technical systems such as electric cars and solar panels. For some, nuclear energy is a solution to the climate crisis, while for others, it is a problem that needs solutions. Solutions take the form of principles or rules of action, such as always trying to take the train instead of flying, or lifestyle choices, such as minimalism and veganism. Solutions can target individual choices and behaviors, such as eco-labels on consumer goods, or they can involve groups, networks, or communities of people, such as "transition towns."

When someone claims that something is a solution, it is often a solution to a problem within the framework of a given system or structure. For the sake of our discussion, we can call this "problem-solving." There are all kinds of problems in human society, and problems are a constant companion to humankind. They have always been a part of social life and will continue to be so. Human beings, and not least scientists, are very creative in framing problems, and in seeking and developing solutions to them. In this sense, the human being is both a problem-framing and a problem-solving creature. Hence, a solution is expected to correct and restore something within the existing structures of society and social life, while transformation involves change on a deeper level. We can, of course, see transformation as a very particular kind of problem-solving, one that involves changing the very form of the structure or system. However, the very word transformation suggests that it must do more than simply restore something to a previous state: it is a change of *form* (see Chapter 6).

Classical sociology focused both on problem solving in the narrow sense and on transformation. For example, many approaches to the study of social problems have a problem-solving character. Many such theories – not least those in the sociological tradition of structural functionalism (Parsons 1937/1968) – focus on how cultural value systems, economic issues, politics, and social relations can be revised in order to restore balance, equilibrium, and harmony to society. Other common forms of sociological analysis have instead critically examined "solutions," critiquing the fact that something was treated as a problem in the first place. They scrutinize instances of power that classify some thing or group as a "problem," as "deviant," and so on, and reveal the various social mechanisms behind such processes of disciplining and stereotyping (e.g., Becker 1963/2018, Foucault

1976/1998, 1977, Goffman 1963/1990). For them, how society defines and frames problems is itself the problem to be critically investigated.

However, other areas of classical sociology also study social transformation. There are macro and universal perspectives, such as Marxist class struggle analysis, historical materialism, and Weber's rationalization process, and there are specific perspectives, such as the study of social movements, one topic of which has been the study of environmental movements (della Porta and Diani 2015). However, the problem-solving mindset is also common in research about social movements and civic action.

To deal with environmental issues, we do need problem-solving capacity, and we are not saying that this mindset is wrong. What we mean is that this way of thinking is inadequate and potentially misleading when applied to our current socio-ecological crisis. We cannot solve a deep crisis with a problem-solving mindset, because such a crisis is related to fundamental structures. Also, if it is impossible to restore a situation or condition, the problem-solving mindset will not help. In environmental issues, the previous state of affairs may be impossible to restore because of irreversible damage (e.g., species extinction, climate change, glacial retreat, desertification, deforestation).

Both the general environmental debate and the scientific and social scientific study of socio-ecological problems and crises have, until recently, tended to adopt this functionalist problem-solving approach, which is based on incremental adaptation to a (gradually) changing environment. If the fundamental problem is how society is organized and functions, then this assumption and worldview must be challenged. There is a need to talk about real transformation and not just address solutions in a narrow sense. Obviously, terms such as "development" and "modernization" that are common in environmental discourse do not solely refer to "returning to a previous state"; they do imply change. Nonetheless, we argue that both public and scholarly debates have generally been about problem solving within the parameters of existing society and social life. It is assumed that (sustainable) development and (ecological) modernization can depend on reforming existing structures (institutions) to deal with serious problems while at the same time achieving "progress." Existing institutions should be reformed to become greener. This is about integrating ecological principles into (reformed) economic and political institutions, but until recently it has only rarely involved examining how existing forms contribute to the reproduction of the problems on a more fundamental level. Instead, there has been a great deal of optimism. There has been a belief that green capitalism and economic growth can be achieved simply by adopting proper pricing mechanisms, introducing and optimizing carbon taxes and climate trading schemes, applying the precautionary principle, developing technological fixes such as negative emission technologies, and so on. To be clear, we are not criticizing carbon taxes, pricing mechanisms, and technological innovation. We are criticizing an overreliance on such instruments that fails to consider transformative change and pay attention to the social. In Box 5.4, we illustrate this narrow mindset with a policy example from Sweden, which expresses the vision that all environmental problems can be solved within a generation.

Box 5.4 The Swedish generational goal

Swedish environmental policy and legislation has generally been guided by the so-called "generational goal." Sweden reformed its environmental legislation in several steps during the 1990s, and in 1999 the Swedish Parliament adopted the generational goal, which is a kind of overarching vision for the legislation. It reads: "The overall goal of Swedish environmental policy is to hand over to the next generation a society in which the major environmental problems in Sweden have been solved, without increasing environmental and health problems outside Sweden's borders" (SEPA 2022). Like the well-known 17 UN sustainable development goals (SDGs) – see Box 5.2 – this overall goal has been broken down into a set of 16 goals in different areas. The division and formulation of these areas generally follows a nature-based type of framework (e.g., clean air, living forests, reduced climate impact, thriving wetlands) with little or no consideration of social aspects. A generation later, all but one of these goals (a protective ozone layer) have failed. The socio-ecological crises of the planet were obviously not possible to solve within one generation. Did the policymakers in one country believe that the problems could be solved within a generation? Was there an over-reliance on technological, cognitive, and structural fixes? Were there social and structural causes behind policy-makers' inability to take into account social and structural causes?

Source: SEPA (2022).

Ten critical questions to reveal inadequate solutions

It is not an easy task to identify and reveal overly simplistic solutions. They are often part of an overall framing of the environmental challenge that makes them seem relevant and effective (see Chapter 4). For example, there are several competing frames of climate change that frame the problem in terms of emissions, market failure, decarbonization, individual consumption, global co-operation, or fossil fuels (Paterson 2021). These different frames open and close particular options, making a proposed solution appear efficient, relevant, and justifiable – while other options are deemed irrelevant, inefficient, or unnecessary. They also either depoliticize or politicize the climate issue, depending on the extent to which they challenge power relations and the current social order. Some frames do not imply any need for deeper changes in society, while others imply a need to restructure society, and thereby challenge prevailing economic and political power structures.

It is therefore a demanding but important task to critically evaluate proposed solutions. This involves asking critical questions about a solution, what it includes and excludes, for whom it will be a solution, and for whom it will make no difference (or even make things worse). As a guide for evaluating proposed solutions, we offer ten critical questions to ask.

1. Is the social taken into account?

This first question follows logically from a key argument of this book. We have strongly emphasized the importance of considering the social, which we view as constitutive of society and therefore always worthy of consideration in the development of environmental policies and solutions. However, when solutions are presented by different actors, they often focus on technologies, market instruments (prices, taxes), legislation, physical planning, and information tools, sometimes together with a discussion of the balance between voluntary and mandatory measures, as well as individual versus collective responsibility (see Chapter 4 on cognitive, technological, and structural fixes). A sociologically informed analysis needs to reflect on how such technological, economic, political, and communicative measures respond to important social matters, such as norms, daily routines, social relationships, social practices, habits of mind in different groups, social stratification, and polarization. All of these kinds of social dynamics must be taken into account when exploring how solutions are understood and applied – appropriated, transformed, ignored, or blocked – or what it is in the everyday lives of individuals and the operations of organizations that is able to translate governance measures and mechanisms into action.

Failure to take the social into account carries a high risk that solutions that look good on paper will later turn out to be either unworkable or to have unintended negative effects, such as causing increased social inequality or fueling political mistrust.

We therefore strongly recommend that proposed solutions take into account the five facets of the social. They must address the inner lives of people in any social group and their thoughts, feelings, and expectations about personal and social life. Solutions must relate to (resonate with, build upon, challenge) existing norms and patterns of social relations and social interactions in the various spheres of life (family, work, school, etc.). Solutions have to relate to (build upon, reconstruct) existing socio-material entities and infrastructures, such as buildings, energy and water supplies, and systems of transportation, provisioning, and communication. Solutions need to address issues of social stratification, for instance how inequality and differences in power and status relate to the reproduction of environmental problems and/or block legitimacy for stronger measures. Solutions need to address the unsustainability of existing institutions as well as involve opportunities for the formation, development, and scaling of new institutions. The following nine questions and recommendations are consistent with this fundamental question.

2. Who is presenting a solution, for what, and for whom?

When someone presents a solution, one must ask who is the presenter and developer of that solution, what is its purpose, and what problem is it intended to address? This is a kind of basic sociological "source criticism." In fact, people often ask critical questions about claim-makers and their supposed hidden motives and agendas. A conspiracy theorist will always be able to find suspicious reasons why someone is presenting a particular solution. However, the sociological source criticism needs to be more sophisticated and nuanced.

A number of follow-up questions can be asked about the problem solver:

- Is the person presenting a solution an expert? What kind of expertise, and in what area? A researcher? Expertise in communication, ecology, economics, policy, or something else?
- Is the person representing an organization? What kind of organization (for-profit, non-profit, public agency) is it, and what is the person's and organization's stake in it?
- What personal (economic, social, and political) interests might the problem solver have?

It should be kept in mind that the presence of additional private incentives other than the publicly stated ones does not preclude the possibility of collective benefits from a proposed solution.

3. Who are the winners and losers?

One needs to ask whose interests, values, and concerns are explicitly or implicitly served by the proposed solution. In almost all cases, a solution will create winners and losers. A power perspective is crucial (see Chapter 2), and the discussions on social stratification and environmental justice in Chapter 3 are also useful for reflecting on this question. When thinking about winners and losers, do not neglect non-human animals and future human generations. Distributional issues are important for reasons of both justice and of legitimacy. The example of the yellow vest protests is instructive in this regard (see Box 3.3). Climate taxes, such as a gasoline tax, will disproportionately affect lower-income groups unless they are compensated in other ways, which may undermine the legitimacy of such solutions. If a solution lacks public legitimacy, it will be very difficult to implement in a democratic society. It is wishful thinking to claim that there will be only winners in working toward a sustainable society. Creating a society that is not dependent on fossil fuels means that the fossil fuel industry has a lot to lose (but it may also have something to gain, if it can rapidly pivot to renewable energy). Solutions with no losers are unlikely to be able to contribute to transformative change, because such change will require sacrifices, particularly among wealthy segments of the population. But one should be especially concerned if the likely losers are found mainly among those who are already vulnerable, marginalized, and discriminated against. This problem of neglecting to identify winners and losers may also be related to the concept of transformation within the sustainability discourse, which Blyth et al. (2019) warn about (see Box 5.5).

Box 5.5 Five risks of the sustainability discourse

As it has become popular to speak in terms of sustainability transformation, there is a risk that the concept will be watered down or used in problematic ways. In their article "The Dark Side of Transformation: Latent Risks in Contemporary Sustainability Discourse," Blyth et al. (2019) address five risks associated with the discourse on sustainability transformation.

Risk 1: Transformation discourse risks shifting the burden of response onto vulnerable parties. For example, citizens and communities are made responsible for building their own resilience, for example in the face of climate change, which is a narrative consistent with neoliberal discourses.

Risk 2: Transformation discourse may be used to justify business-as-usual. Linked to framings around the green economy, ecosystem services, green growth, and the wider adoption of green market language, transformation may become a catch-all term. It may be associated more with adaptation and building resilience in the existing system rather than transformation.

Risk 3: Transformation discourse pays insufficient attention to social differentiation. The warning here is that the discourse may be characterized by a consensus-oriented win-win rhetoric, and fail to consider the winners and losers of proposed solutions, and how this relates to social stratification.

Risk 4: Transformation discourse can exclude the possibility of non-transformation or resistance. It is argued that it is important not to lose sight of those who want to resist transformation proposals coming from powerful actors (e.g., investment in renewable energy), because these proposals may neglect local concerns.

Risk 5: Insufficient treatment of power and politics threatens the legitimacy of transformation discourse. The literature tends to be optimistic, but often fails to consider the political processes that underpin transformation and how power dynamics are involved.

Hence, a key thing to understand is that the concept of transformation itself can be loaded with power, and may be co-opted by powerful elites. It is important to acknowledge the role of power and politics, as well as to adopt a pluralistic understanding of transformation, Blyth et al. argue.

4. Are there hidden compromises?

It is often the case that a solution reflects compromises between different interest groups. Sociological analysis helps to make hidden processes of negotiation more transparent. We can illustrate this with the example of eco-labels. These are seen as a consumer-oriented solution, as facilitating consumer power or political consumerism in the form of "buycotting" (Boström et al. 2019). What is often hidden behind the label is that the principles, criteria, and indicators in an eco-labeling scheme are the result of social interaction, or more precisely, negotiations between actors in the field, including business actors, civil society organizations, and other environmental experts (Boström and Klintman 2008). These groups have their different motives for participating in joint efforts to introduce an eco-labeling scheme to the market. Such schemes are therefore based on compromises, while they are often represented to the public as "objective," "neutral," "expert-based," or simply "good for the environment" (Boström and Klintman 2008, 2019a). The problem is not that these schemes rely on compromises. If they did not, they would be hard to implement in a market. The problem is that if consumers believe the act of buying eco-labeled goods can literally be considered an act of "going green," they may be misled into thinking that the problems associated with consumption can be solved simply by consuming products with the right labels. Such a belief can lead to problems known as "ethical fetishism," "wasteful distraction," "moral licensing," and various kinds of compensatory thinking (see Boström 2023, Boström and Klintman 2019a, 2019b). Thus, solutions can be deceptive, which we discuss in question 6 below.

5. Does it address symptoms or causes?

Is a solution a way of adapting to a problem – treating its symptoms, like a painkiller – or is it a way of alleviating the problem, that is, removing its root causes? If the problem is overconsumption, buying eco-labeled goods is certainly not the solution, given the rebound effect discussed above. A proposal that appears to be a solution to a particular individual or community may not be a solution at the aggregate level. The group proposing it may simply be adapting to the existence of problems with treatments relevant to their specific situation. Of course, people and communities need to build capacity in the face of increased risks from hurricanes, wildfires, and sea level rise. But the capacity to build such resilience varies widely both between and within countries. Stratified adaptive capacity is an issue of environmental justice (see Chapter 3). The rich may be able to move to new locations to avoid pollution, noise, and vulnerability to climate change (Kenner 2019). If the option of adaptation becomes feasible and attractive to wealthy populations, it may reduce incentives and capacities for mitigation, thereby increasing the problems for poorer segments of society.

We can also look at "corporate sustainability" or "corporate social responsibility" (CSR) and suggest that analysts studying CSR programs ask whether these are ways of addressing symptoms rather than root causes. One should look for problematic circumstances in the production context. CSR generally fails to address many of the root causes of environmental and ethical problems, and it has generally been adopted by large multinational corporations that have outsourced production to countries with lax environmental regulations, poor working conditions, and low wages. CSR is not about counteracting this drive to cut costs, because the drive itself is structurally related to global market competition (see, e.g., Locke 2013). Instead, the solutions associated with it, such as green brands, eco-labels, and codes of conduct that are applied to "responsible" supply chain management, environmental management systems, sustainability reporting, transparency, sustainability communication, and so on, are mostly about "impression management." While CSR programs may eliminate some of the most harmful impacts and sweatshop working conditions, they reflect an effort to create social legitimacy for one's business – a license to operate – rather than to achieve societal change or truly engage in more responsible business practices. CSR is not about sharing economic power and wealth more equitably. While the types of solutions we list above may not be inherently wrong (on the contrary, they may be important), the problem has to do with the structural inability of companies operating in a competitive capitalist market to actually grapple with root causes (see Chapter 2). Companies may discuss climate change and other global environmental problems in a serious way, cite trustworthy sources such as the IPCC, adopt "eco-efficient technologies," implement codes of responsible conduct for themselves and their suppliers, and offer "green products" for the "green market," yet fail to engage in any self-critical discussion about how their operations contribute to asymmetrical global power relations and a global and local order of enormous environmental impact (see Dauvergne and Lister 2013, Brand and Wissen 2021).

6. Does it distract or mislead?

The above point about corporate social responsibility can be expanded by asking another question. Is the presented solution effective, or is it just rhetoric that gives the appearance of being responsible and committed action? The main problem may not be that actors are making PR out of their own "solutions." The problem may be that the solution distracts us and diverts attention away from more problematic circumstances (i.e., root causes), such as people's and organizations' overconsumption or their globally/locally exploitative business practices. Creating distractions need not be a deliberate and cynical strategy. Honest "problem-solvers" may unknowingly mislead themselves and others about the effectiveness of a particular solution, believing that it is a way to be environmentally responsible. As we discussed in Chapter 4, framings and storylines make us see particular solutions and fixes as relevant and effective. Thus, an organization or individual can believe in a proposed solution that is inadequate in practice.

A common form of window dressing and greenwashing that corporations engage in is to present themselves and their green products as the solution to a problem, while ignoring how their business contributes to the environmental crisis more broadly. Similarly, climate policies are often presented by governments as a means of creating "green growth," that is, of creating new markets and new employment opportunities. A radical climate policy may imply the growth of some activities and services, but this should not overshadow the fact that it will also require changes in current levels and patterns of production, consumption, and transport.

Illustrative examples can be found in popular systems of climate compensation. The rhetoric of "net-zero emissions," "climate neutrality," and "climate-positive technologies" is generally misleading. Net-zero carbon emissions – using carbon offsets to compensate for carbon emissions – is essential to limiting global warming. The problem is that it is often misused by actors who fail to link it to a long-term strategy for limiting global emissions. An analysis of net-zero carbon emissions shows that it needs to be used in cautious, comprehensive, and effective ways, and when it is used it needs to be part of broader socio-ecological objectives that also include issues of fairness (Fankhauser et al. 2022). In practice, this is rarely the case. Climate offsetting schemes generally have several problematic features, one of which is that they are often applied without any intention of reducing climate emissions.[3] These phrases, commonly used in green marketing, give the dubious impression that something like a "climate neutral" hamburger has no carbon footprint at all. Can you even do something good for the environment by driving an electric SUV? These schemes are often used to legitimize business as usual, continued luxury consumption, and mass consumption.

7. Is there a gap between statement and action?

The so-called attitude-behavior gap, or value-action gap, is a common topic in sustainable consumption research (see Chapter 2). A gap between stated goals and actions taken can appear among both individuals and collective actors such as countries and organizations. Obviously, there are plenty of examples where no real action is intended, only green talk, or where there is a commitment to green action, but it is not strong enough to override other goals (such as material wealth, social status, or market growth). There are also examples where people and organizations sincerely want to change their actions, but there are many factors that prevent such change. Boström and Klintman (2019a, 2019b; see also Boström 2023) list a number of factors that hinder sustainable consumption or reduced consumption:

- Unable to afford more expensive green products.
- Social dilemmas: "why should I pay a more for green products when most other consumers do not?"
- Motivational complexities: actors have competing goals, values, and priorities.
- Lack of information, trust, and perceived effectiveness of the available "green" options.
- Lock-in of structural arrangements that cause path-dependencies of action, and a lack of practical arrangements that can facilitate climate-conscious or other environmentally mindful choices.
- Constant pressure from mass consumption cultures and policy, which is difficult to resist.
- Consumerist norms and habits of social interaction that prevent consumers from reducing their consumption or choosing alternative forms of consumption: the fear of being stigmatized or harassed, the difficulty of finding other ways of socializing than through shopping, for example.
- Unconscious consumerism: deeply socialized, naturalized/normalized, and habitualized consumption.

So, there are a myriad of factors that make it difficult for actors to do what they actually want to do. To tie this in with our earlier analysis, we can say that actors are committed on a cognitive, emotional, or value level, but are paralyzed by the complexity of society and their embeddedness in various social circumstances, such as social relationships.

8. Here or there, now or later?

A crucial feature of many environmental problems is that they are dispersed in time and space; activities in one place can impact other places and people. This is the topic of Chapter 3, that environmental problems are spread – spatially, temporally, and socially. It is also a central concern of climate and environmental justice to explore and expose how activities far away have serious impacts on the lives of human and non-human beings. For this reason, it is always important to ask whether a proposed solution actually "solves" an environmental problem "here" and "now," but does so by shifting the problem to other places and regions ("there") or by postponing the consequences ("later"). This is partly touched upon in point three above, "Who are the winners and losers?" But here we extend it to looking beyond a particular society. Asking questions about the temporal and spatial consequences of a proposed solution means discussing a proposed solution in a broader perspective. In Chapter 3, we used the example of electric vehicles, a solution that has been proposed by many stakeholders as a way to achieve climate-neutral transportation. However, this solution is based on the mining of precious minerals, which is done in environmentally harmful ways.

Thus, what is a green solution in one place may in fact be harmful to the environment somewhere else. To this we can add that the electrification of transportation implies a dramatic increase in electricity consumption. Countries might, for example, build nuclear power plants to meet this increased demand, and while this is much more climate-friendly than oil and coal-based electricity production, it creates radioactive waste (in the form of spent nuclear fuel, for example) that must be stored and kept separate from human activities for 10,000 years. If leakage occurs from these repositories, it will mean that nuclear power has created "clean" energy for today's society while passing on risks and dangers to future generations and societies. What seems like a solution now may prove to be anything but environmentally friendly in the future. Similarly, negative emissions technologies (NETs), such as geoengineering and carbon capture, are presented as important ways to limit global warming to 1.5°C (IPCC 2018). Even if NETs prove to be an environmentally sound and cost-effective way to limit global warming, [4] they may have considerable social impacts, including changes in land rights, displacement of people, and competition for food supplies (Beck and Mahony 2018). Similarly, climate compensation or carbon offset schemes – where individuals, companies, organizations, and even whole countries finance greenhouse gas savings elsewhere and then register this reduction as their own – mean that an environmental emission "here" is compensated by a reduction "there" (see our discussion in question 6 above). But this reduction may have major negative social and environmental impacts on the people and non-human animals living in that other place. Compensating for air travel by planting trees in another country means that another country – often in the global South – has to change its land use to make space for planting trees. It is therefore always important to explore and think about how a solution might actually worsen the social and environmental conditions elsewhere – for other groups, places, generations, and animals. It is also important to ask whether this is a sustainable solution in the long run. If offsets are used as a way to legitimize continued and increased carbon dioxide emissions, then only the symptoms are being treated, while the root causes are not being addressed at all.

9. Will there be unintended effects?

Unintended consequences frequently accompany human actions. Environmental problems are a good example of this; they are not aimed for, and to a large extent are unanticipated. The macro-sociological theories presented in Chapter 2 – the treadmill of production, the risk society, ecological modernization, and postcolonialism/political ecology – all view environmental problems as the downside of human planning and technological development, but propose different ways of handling them. Environmental solutions also regularly lead to unintended consequences. An example is ozone depletion (see Box 4.3, Chapter 4). The class of chemical compounds known as CFCs was developed in the late 1920s and was welcomed by industry. Unlike other substances used in refrigerators, CFCs were not toxic or flammable. They became widely used in society, as refrigerants, propellants (in spray bottles), and solvents. In the early 1970s, it was discovered that they had destructive effects on the stratospheric ozone layer. They were gradually regulated and later banned, being phased out and replaced by other chemical compounds. Thus, what was seen as a solution – a non-toxic compound – was later found to cause severe environmental problems.

This applies not only to specific chemical substances and technologies, but to all kinds of human activity. Current calls to restructure economic sectors – such as agriculture, transportation, and energy – can help create a more sustainable society, but at the same time it is necessary to undertake the difficult task of exploring and reflecting on the possible unintended and negative consequences of such proposals. This is especially important when discussing large-scale changes. In his book *Seeing Like a State* (1998), James Scott explores how well-intentioned, grand utopian schemes to improve society have brought death and disruption. State-led projects such as the Great Leap Forward in China, agricultural “modernization” in the tropics, scientific forestry in Germany, and forced resettlement in Tanzania, have all had serious unintended consequences. The reason for this is that the actors (including states) initiating these large-scale plans have an oversimplified view of reality that does not do justice to the complexity of social life and ecological systems.

Box 5.6 Robert Merton and unanticipated consequences

The first systematic sociological study of unintended consequences was made by the US sociologist Robert Merton in 1936 when he published his classic paper "The Unanticipated Consequences of Purposive Social Action." He identified four causes of such consequences: (i) Ignorance: the actor lacks the time and energy to reflect over possible outcomes of an action; (ii) Error: the actor has a distorted understanding of the situation and the probable chain of events; (iii) The imperious immediacy of interest: the actor sees only the short-term, intended consequences, and not the later or different consequences of the action taken; (iv) Self-defeating predictions: predictions about human actions can become new elements in the concrete situation, and thus change the outcome of the original course of action.

In analyzing unanticipated consequences, Merton also raises two methodological problems. The first is the question of causal imputation: What consequences can justifiably be attributed to a certain action? The second is the problem of ascertaining the actual purpose of a given action. Judging the reliability of post facto declarations of intent is notoriously complicated. However, as Merton states, it is less difficult to trace intentions with formal organizations than with individuals, because organizations have explicit procedures and statements of purpose.

Later contributions have highlighted the conflation of unanticipated and unintended consequences, showing that much of the literature today treats them as synonyms (Baert 1991, Boudon 1982, Zwart 2015). Many unintended consequences are known in advance (anticipated) but are not intended. An actor may see them as possible but unlikely outcomes, or as legitimate side effects (if the intended outcome is considered more important than the unintended ones) (Lidskog and Sjödin 2018).

Of course, it is not possible to completely avoid unintended consequences, but it is possible to anticipate and prevent at least some of them. It is important to strive to understand the limits of knowledge, to be cognizant of what is not known, and to take this into account in planning and decision making. Thus "active non-knowledge" – making an effort to reveal areas of non-knowledge and make them visible – is important because it makes us more humble and sensitive, and thus better able to adapt to new circumstances (Gross 2007, 2022). When evaluating a proposed "solution," it is therefore always important to critically reflect on whether possible unintended effects have been taken into account. To what extent have the limits of knowledge been considered? What is not known? What is uncertain? What knowledge has been excluded? And finally, are the facets of the social taken into account, or is the analysis limited to technical or economic solutions, without consideration of the fact that they have to be implemented in a stratified society?

10. The way forward: Small steps or a giant leap?

Much of the analysis in this chapter, including resistance to change, the perception of smallness in relation to the complexity of global society, and the adoption of a problem-solving mindset, suggests that the typical "way forward" is not one of incremental change achieved through "small steps." Many of the solutions we have discussed are not even steps forward, but rather steps along a path that goes around in circles, constantly sustaining the unsustainable.

We are not suggesting, however, that taking small steps is necessarily wrong. Viewing change as a stepwise process can be a sound approach. In fact, many radical social changes not only begin with small steps, but are also constituted by them. It is a matter of trial and error, of experimentation, of seeing if something new works in the social context. If the steps can be envisioned to be part of a longer walk toward change, involving collective efforts on a much larger scale, then these steps may be wise. A "small" step need not be trivial if it is considered to be part of a process leading toward large-scale transformative change. Many "trivial" actions can add up to significant change. The theory of leverage points states that there are places in a complex system where small steps, resulting in small shifts, can produce system changes through leverage mechanisms (Meadows 1999, 2008). An entire society can begin to change through interventions that make changes in taxes and subsidies, information flows, incentive structures, and organizational goals. This requires that these small steps be linked to mechanisms that produce large-scale change.

We also need to remember that rhetoric about "giant leaps" (or similar) can be unrealistic and bombastic. Phrases such as "overthrow the system now" can sound naive, are unlikely to attract sufficient public support, and run the risk of leading to disappointment or unpleasant outcomes – both environmental, as discussed by Scott (1998) (see question 9 above) and political (such as restrictions on democracy and violations of human rights).

Some change processes may need to be gradual. Others may be more immediate, radical, and far-reaching, depending on available opportunities, resources, and ability to mobilize. Still others will be triggered by external events such as a pandemic or a disaster. So we remain open-minded and undecided on this last question. We suggest that the analyst needs to be cautious and reflective. The important thing is to determine whether the proposed "steps" are actually linked to major transformative change.

Conclusion

In this chapter we have discussed the need to apply an attitude of constructive criticism to the development and presentation of solutions. We have argued that many solutions are better understood as barriers to change than as catalysts for change. Many proposals that are presented as solutions are in fact best understood as contributing to societal inertia rather than being a part of transformative change. Barriers need to be understood in terms of resistance to change and the complexity of society, as well as their involvement with strong power structures. Resistance to change can be both active and passive, intentional and unintentional, conscious and unconscious. This chapter has shown how complexity and resistance to change relate to the five facets of the social. We have also shown that attempts to change society should not be designed from within a narrow problem-solving mindset. Our ten questions and accompanying recommendations are intended to equip the analyst with the critical and constructive gaze and sociological competence needed to evaluate proposed solutions in a cautious and nuanced way that avoids black-and-white thinking.

Therefore, the key word we suggest is not solution but transformation. This is the focus of our concluding chapter. We must dare to speak of the need for transformation in a quite literal sense. If we face global problems of a potentially catastrophic nature that can cause irreparable damage to life on Earth, we should not (only) speak of problem-solving. We should not engage in wishful thinking. We should talk about transformation. Of course, we can speak of a variety of "solutions," or a "portfolio of solutions," within a general and more holistic framework of "transformation." Society will be transformed in the long run, no matter what, regardless of whether or not humanity tries to make it happen. If the crisis is not averted with sufficient transformative action, the crisis itself will bring about a transformation. This disaster-driven transformation, fringed by droughts, floods, heat waves, and wildfires, is much more unpleasant to contemplate than beginning the work of transforming society today, and thereby avoiding many catastrophes.

Questions for reflection and discussion

This chapter presents ten questions to ask when reviewing proposals for environmental solutions. We also suggest asking questions about barriers to change, complexity, and resistance. We invite you to use these concepts to examine your local context: the household, university, or organization where you work or spend your free time, the community where you live, and the government of your country.

The authors discuss how complexity can cause social paralysis. Can you find examples of actors citing complexity, or similar reasons, as an excuse for inaction? How could these actors be motivated to take action?

The authors discuss the impossibility of getting a complete overview of today's complex society, but what could be a useful approach to get at least a partial overview or understanding of the relationship between society and the environment? Can the 17 SDGs of Agenda 2030 be helpful? (See Box 5.1.) Can you find any important issues that these 17 goals do not cover?

Look at Box 5.2. Do you or others in your social network and community use any of the justifications listed? How can resistance to change be dealt with?

The authors discuss the need to distinguish between piecemeal solutions within a problem-solving mindset and solutions with transformative potential. Do you agree that it is possible to make such a distinction? Where would you draw the line?

Before turning the page to the next and final chapter, you may want to take a moment to reflect on how and where in society you see opportunities for transformative change.

Notes

1 The "Jevons Paradox" states that increased efficiency makes a product or activity cheaper, and this tends to lead to increased consumption. It was first articulated by William Stanley Jevons in his book *The Coal Question* (1865/2001), where he observed that technological improvements in the use of coal in producing iron led to increased, not decreased, consumption of coal. This is the first formulation of the "rebound effect" (see York and McGee 2016, for a discussion of the Jevons Paradox).

2 In the case of ozone depletion, even though CFC substitutes have shorter lifetimes and lower ozone depletion potentials, they are not without environmental impacts. In the case of transboundary air pollution, there are other airborne substances besides sulfur that need to be substantially reduced, such as particulate matter and nitrogen dioxides (Lidskog and Sundqvist 2011).

3 Another problem concerns the permanency and additionality of climate compensation. If a company (or citizen) plants a tree to compensate for carbon emissions, this presupposes that the tree would not have been planted without the money invested in it, which is very difficult to prove. Moreover, this planting of trees may be used by other actors to legitimize deforestation in other places, which means that on a global level, there is no compensation.

4 For a critical discussion of bioenergy with carbon capture and storage (BECCS), see Haikola et al. (2021).

References

Baert, P. (1991) 'Unintended consequences: A typology and examples', *International Sociology*, 6(2), pp. 201–210. https://doi.org/10.1177/026858091006002006

Beck, S. and Mahony, M. (2018) 'The IPCC and the new map of science and politics', *WIREs Climate Change*, 9(6), e547. https://doi.org/10.1002/wcc.547

Beck, U. (1992) *Risk society: Towards a new modernity*. London: Sage.

Beck, U. (2000) *What is globalization?* Malden: Polity Press.

Beck, U. (2009) *World at risk.* Cambridge, UK: Polity.

Becker, H. (2018) *Outsiders: Studies in the sociology of deviance.* New York: Free Press [originally published 1963].

Berger, P.L. and Luckmann, T. (1991) *The social construction of reality: A treatise in the sociology of knowledge.* London: Penguin [originally published 1967].

Blühdorn, I. (2011) 'The politics of unsustainability: COP15, post-ecologism and the ecological paradox', *Organization & Environment*, 24(1), pp. 34–53. https://doi.org/10.1177/1086026611402008

Blyth, J., Silver, J., Evans, L., Armitage, D., Bennett, N.J., Moore, M-L., Morrison, T.H., and Brown, K. (2019) 'The dark side of transformation: Latent risks in contemporary sustainability discourse', *Antipode*, 50(5), pp. 1206–1223. https://doi.org/10.1111/anti.12405

Boltanski, L. and Thévenot, L. (2006) *On justification: Economies of worth.* Princeton: Princeton University Press.

Börjeson, N. and Boström, M. (2018) 'Towards reflexive responsibility in a textile supply chain', *Business Strategy and the Environment*, 27(2), pp. 230–239. https://doi.org/10.1002/bse.2012

Boström, M. (2015) 'Between monitoring and trust: Commitment to extended upstream responsibility', *Journal of Business Ethics*, 131(1), pp. 239–255. https://doi.org/10.1007/s10551-014-2277-6

Boström, M. (2023) *The social life of unsustainable mass consumption.* Lanham: Lexington Books.

Boström, M. and Klintman, M. (2008) *Eco-standards, product labelling, and green consumerism.* Basingstoke: Palgrave Macmillan.

Boström, M. and Klintman, M. (2019a) 'Can we rely on "climate friendly" consumption?', *Journal of Consumer Culture*, 19(3), pp. 359–378. https://doi.org/10.1177/1469540517717782

Boström, M. and Klintman, M. (2019b) 'Mass-consumption and political consumerism', in M. Boström, M. Micheletti and P. Oosterveer (eds.) *The Oxford handbook of political consumerism.* Oxford University Press, pp. 855–878.

Boström, M., Micheletti, M. and Oosterveer, P. (2019) 'Studying political consumerism' in M. Boström, M. Micheletti and P. Oosterveer (eds.) *Oxford international handbook of political consumerism.* Oxford University Press, pp. 1–24.

Boudon, R. (1982) *The unintended consequences of social action.* London: Macmillan.

Brand, U. and Wissen, M. (2021) *The imperial mode of living. Everyday life and the ecological crisis of capitalism.* London: Verso.

Brenner, N. (2004) *New state spaces: Urban governance and the rescaling of statehood.* Oxford: Oxford University Press.

Bullard, R.D. (2000) *Dumping in dixie: Race, class and environmental quality.* 3rd edn. Boulder, CO.: Westview.

Cialdini, R.B. (2007) 'Descriptive social norms as underappreciated sources of social control', *Psychometrika*, 72, pp. 263–268. https://doi.org/10.1007/s11336-006-1560-6

Dauvergne, P. and Lister, J. (2013) *Eco-business: A big-brand takeover of sustainability.* Cambridge, MA: MIT Press.

della Porta, D. and Diani, M. (eds.) (2015) *The Oxford handbook of social movements.* Oxford: Oxford University Press.

Durkheim, É. (2013). *The division of labour in society.* 2nd edn. London: Palgrave Macmillan [originally published 1893].

Eyal, G. (2019) *The crisis of expertise.* Cambridge: Polity Press.

Fankhauser, S., Smith, S.M, Allen, M., Axelson, K., Hale, T., Hepburn, C., Kendal, M.K., Koshla, R. et al. (2022) 'The meaning of net zero and how to get it right', *Nature Climate Change*, 12, pp. 15–21. https://doi.org/10.1038/s41558-021-01245-w

Foucault, M. (1977) *Discipline and punish: The birth of the prison.* New York: Pantheon.

Foucault, M. (1998) *The will to knowledge: The history of sexuality: 1.* London: Penguin books [originally published 1976].

Fridell, G. (2019) 'Conceptualizing political consumerism as part of the global value chain' in M. Boström, M. Micheletti and P. Oosterveer (eds.) *The Oxford handbook of political consumerism.* Oxford University Press, pp. 249–274.

Giddens, A. (1990) *Modernity and self-identity: Self and society in the late modern age.* California: Stanford University Press.

Goffman, E. (1990) *Stigma. Notes on the management of spoiled identity.* New Jersey: Penguin books [originally published 1963].

Gross, M. (2007) 'The unknown in process: Dynamic connections of ignorance, non-knowledge and related concepts', *Current Sociology*, 55(5), pp. 742–59. https://doi.org/10.1177/0011392107079928

Gross, M. (2022) 'Technological fixes: Nonknowledge transfer and the risk of ignorance', in L. Pellizzoni, E. Leonardi and V. Asara, (eds.) *Handbook of critical environmental politics.* Cheltenham, UK: Edward Elgar, pp. 308–317.

Gunderson, R. (2022) 'Powerless, stupefied, and repressed actors cannot challenge climate change: Real helplessness as barrier between environmental concern and action', *Journal for the Theory of Social Behaviour*, 53(2), pp. 271–295. https://doi.org/10.1111/jtsb.12366

Haikola, S., Anshelm, J. and Hansson, A. (2021) 'Limits to climate action: Narratives of bioenergy with carbon capture and storage', *Political Geography*, 88, 102416. https://doi.org/10.1016/j.polgeo.2021.102416

Hansen, A. (2023) 'Capitalism, consumption, and the transformation of everyday life: The political economy of social practices', in A. Hansen and K. Nielsen (eds.) *Consumption, sustainability, and everyday life.* Cham: Palgrave Macmillan, pp. 27–54. https://doi.org/10.1007/978-3-031-11069-6_2

Head, B.W. and Alford, J. (2015) 'Wicked problems implications for public policy and management', *Administration & Society*, 47(6), pp. 711–739. https://doi.org/10.1177/0095399713481601

Held, D. (ed.) (2004) *A globalizing world? Culture, economics, politics*. 2nd edn. London: Routledge.

Höijer, B., Lidskog, R. and Uggla, Y. (2006) 'Facing dilemmas. Sense-making and decision-making in late modernity', *Futures*, 38(3), pp. 350–366. https://doi.org/10.1016/j.futures.2005.07.007

IPCC (2018) *Global warming of 1.5 C: An IPCC special report on the impacts of global warming of 1.5 C above pre-industrial levels and related global greenhouse gas emission pathways, in the context of strengthening the global response to the threat of climate change, sustainable development, and efforts to eradicate poverty.* Geneva: IPCC. https://www.ipcc.ch/sr15/

Jackson, T. (2017) *Prosperity without growth: Foundations for the economy of tomorrow.* 2nd edn. London: Routledge.

Jevons, W.S. (2001) *The coal question: An inquiry concerning the progress of the nation, and the probable exhaustion of our coal-mines* (Palgrave archive edn.) Basingstoke: Palgrave [originally published 1865].

Kenner, D. (2019) *Carbon inequality: The role of the richest in climate change.* London: Earthscan.

Klintman, M. (2019) *Knowledge resistance: How we avoid insight from others*. Manchester: Manchester University Press.

Leonard, L. (2009) *Civil society reflexiveness in an industrial risk society.* PhD diss., University of London (Kings College).

Leonard, L. and Lidskog, R. (2021) 'Industrial scientific expertise and civil society engagement: Reflexive scientisation in the South Durban Industrial Basin, South Africa', *Journal of Risk Research*, 24(9), pp. 1127–1140. https://doi.org/10.1080/13669877.2020.1805638

Lidskog, R. and Elander, I. (2010) 'Addressing climate change democratically: Multi-level governance, transnational networks and governmental structures', *Sustainable Development*, 18(1), pp. 32–41. https://doi.org/10.1002/sd.395

Lidskog, R. and Sjödin, D. (2018) 'Unintended consequences and risk(y) thinking: The shaping of consequences and responsibilities in relation to environmental disasters', *Sustainability*, 10(8), 2906. https://doi.org/10.3390/su10082906

Lidskog, R. and Sundqvist, G. (2011) 'Transboundary air pollution policy in transition', in R. Lidskog and G. Sundqvist (eds.) *Governing the air: The dynamics of science, policy, and citizen interaction.* Cambridge, Ma.: MIT Press, pp. 1–36.

Lidskog, R. and Sundqvist, G. (2018) 'Environmental expertise', in M. Boström and D. Davidson (eds.) *Environment and society: Concepts and challenges*. Basingstoke: Palgrave, pp. 167–186.

Lidskog, R. and Sundqvist, G. (2022) 'Lost in transformation: The Paris Agreement, the IPCC, and the quest for national transformative change', *Frontiers in Climate*, 4, 906054. https://doi.org/10.3389/fclim.2022.906054

Lidskog, R., Bishop, K., Eklöf, K., Ring, E., Åkerblom, S., and Sandström, C. (2018) 'From wicked problem to governable entity? The effects of forestry on mercury in aquatic ecosystems', *Forest Policy and Economics*, 90, pp. 90–96. https://doi.org/10.1016/j.forpol.2018.02.001

Locke, R.M. (2013) *The promise and limits of private power.* New York: Cambridge University Press.

Meadows, D. (1999) *Leverage points: Places to intervene in a system*. Hatland, VT: The Sustainability Institute. https://donellameadows.org/wp-content/userfiles/Leverage_Points.pdf

Meadows, D.H. (2008) *Thinking in systems: A primer*. Vermont: Chelsea Green Pub.

Merton, R.K. (1936) 'The unanticipated consequences of purposive social action', *American Sociological Review*, 1(1), pp. 894–904 [reprinted as 'Unanticipated consequences of social action' in Merton, R.K. (1976) *Sociological Ambivalence and Other Essays.* New York: The Free Press, pp. 145–155].

Miller, D. (2010) *Stuff.* Cambridge: Polity Press.

Norgaard, K.M. (2011) *Living in denial: Climate change, emotions, and everyday life.* Cambridge, Mass.: MIT Press.

Oreskes, N. and Conway, E.M. (2010) *Merchants of doubt: How a handful of scientists obscured the truth on issues from tobacco smoke to global warming.* New York, NY: Bloomsbury Press.

Parsons, T. (1968) *The structure of social action: A study in social theory with special reference to a group of recent European writers.* 2nd edn. New York: Free Press [originally published 1937].

Paterson, M. (2021) '"The end of the fossil fuel age"? Discourse politics and climate change political economy', *New Political Economy*, 26(6), pp. 923–936. https://doi.org/10.1080/13563467.2020.1810218

Rinkinen, J., Shove, E. and Marsden, G. (2021) *Conceptualizing demand: A distinctive approach to consumption and practice.* London: Routledge.

Rittel, H.W. and Webber, M.M. (1973) 'Dilemmas in a general theory of planning', *Policy Sciences*, 4, pp. 155–69. https://doi.org/10.1007/BF01405730

Scott, J.C. (1998) *Seeing like a state: How certain schemes to improve the human condition have failed.* New Haven: Yale University Press.

SEPA 2022 *Environmental objectives.* https://www.naturvardsverket.se/en/environmental-work/environmental-objectives/ (accessed 29 November 2022).

Shove, E. (2003) *Comfort, cleanliness and convenience: The social organization of normality.* Oxford: Berg.Smith, A. (2008). *An inquiry into the nature and causes of the wealth of nations.* Oxford: Oxford University Press [originally published 1776)

Speth, J.G. and Haas, P.M. (2006) *Global environmental governance.* Washington: Island Press.

Stoddard, I., Anderson, K., Capstick, S., Carton, W., Depledge, J., Facer, K., Gough, C., Hache, F., et al. (2021) 'Three decades of climate mitigation: Why haven't we bent the global emissions curve?', *Annual Review of Environment and Resources*, 46, pp. 653–689. https://doi.org/10.1146/annurev-environ-012220-011104

UN (2022) *The Sustainable Development Goals Report 2022.* New York: United Nations Publications ISBN 978–92-1–101448-8. https://unstats.un.org/sdgs/report/2022/

UN (n.d.) *Sustainable development*. https://sdgs.un.org

Wallerstein, I.M. (2004) *World-systems analysis: An introduction*. Durham: Duke University Press.

Wilkinson, R. and Pickett, K. (2011) *The spirit level: Why greater equality makes societies stronger.* London: Bloomsbury.

Wilkinson, R. and Pickett, K. (2018) *The inner level: How more equal societies reduce stress, restore sanity and improve everybody's well-being.* London: Allen Lane.

World Bank (2018) *Overcoming poverty and inequality in South Africa: An assessment of drivers, constraints and opportunities.* Washington, DC: The World Bank. http://documents.worldbank.org/curated/en/530481521735906534/pdf/124521-REV-OUO-South-Africa-Poverty-and-Inequality-Assessment-Report-2018-FINAL-WEB.pdf (accessed 29 November 2022).

York, R. and McGee, J.A. (2016) 'Understanding the Jevons Paradox', *Environmental Sociology*, 2(1), pp. 77–87. https://doi.org/10.1080/23251042.2015.1106060

Zwart, F. (2015) 'Unintended but not unanticipated consequences', *Theory and Society*, 44, pp. 283–297. https://doi.org/10.1007/s11186-015-9247-6

6 Transformation

Ways of changing society

A core message of this book is that we need to discuss transformation rather than solutions. We need to change not just a particular sector or area of activity, but society as a whole. Transformation also means rethinking the way we bring about change. A major problem is the tendency to see practices, norms, and institutions as fixed and almost unchangeable entities. But, as emphasized in previous chapters, what seem to be permanent features of society can gradually or suddenly dissolve. Empires (e.g., Rome), political and economic systems (e.g., feudalism), and nation-states (e.g., Yugoslavia) may cease to exist. Thus, changing society does not only mean initiating and implementing change through current organizations and institutions and their usual tools. It also means being creative in inventing new ways to initiate and facilitate change. Nothing is too small or too large to be changeable, and achieving a sustainable society will require that all five facets of the social are included in this work, transforming everything from our inner lives to global institutions.

By highlighting barriers such as complexity, resistance, and inadequate solutions in the previous chapter, the last thing we wanted to do was create a sense of impossibilism (Jackson 2017), which could lead to resignation and paralysis. It is important to remember that throughout history it has often been claimed that existing circumstances cannot be changed. There have been messages about the impossibility of abolishing slavery, colonialism, and apartheid, and of creating democracy and spreading human rights. The language of impossibilism is the rhetoric of the powerful and privileged, and, according to Armstrong (2019), we are often wrong about what is possible. We may believe that something is impossible, a belief we can maintain simply because we do not want to try to change society. Indeed, many countries have lifted many people out of poverty. Welfare societies have been created. Human rights and democracy have been established and institutionalized in many countries. Moreover, the feasibility of transformative change should not be seen as a black-and-white issue. We can abandon unsustainable practices, even if it will take a long time to develop a sustainable society. At the same time, we need to recognize the barriers identified in Chapter 5 and avoid wishful thinking. From a sociological perspective, social processes combine elements of determinism and voluntarism; change is possible, but it often requires persistent struggle and is conditioned by existing circumstances. The sociologist Anthony Giddens (1990) spoke

DOI: 10.4324/9781032628189-6

of "utopian realism," which highlights the importance of visions that are not based on wishful thinking, but are possible, if difficult, to achieve. Similarly, in this book we have argued against impossibilism, cynicism, and defeatism, while at the same time avoiding naive views of change. Instead, we can speak of a *realistic possibilism*, which means that transformation is one of many possibilities, and realizing it requires a firm attention to existing societal conditions.

Transforming society means not only developing sustainable activities, but also phasing out unsustainable ones (see Figure 6.1). For example, it is not enough to increase the use of renewable energy if fossil fuels are not phased out at the same time. By identifying three phases of transformation, the figure stresses the need for experimentation, testing, innovation, and learning. This must be followed by broad and rapid sharing, dissemination, and support of sustainable ways of organizing society and social life. Finally, this new organization must be institutionalized and become the new normal.

Ours is a time of contradictions. We live in a world with old structures that continually contribute to unsustainable practices and need to be gradually replaced by new ones. This means that we cannot immediately transition from society A to a desired society B (and no one knows what society B will look like), because it is a matter of many parallel processes that influence each other. This is a situation where we are trying to reform the existing society to make it less unsustainable while at the same time making space for new and more sustainable ways of living and organizing our society. This process must not proceed too slowly either, given the urgency of the environmental crisis. As the IPCC states in its latest assessment report (IPCC 2023), this decade will require immediate, rapid, and greatly reduced greenhouse gas emissions from all sectors. Research, negotiations, agreements, and policies are important and necessary, but if such desk work is not followed by practical action and changes in activities, it will not matter. So we are in a challenging situation that requires both deliberation and action. We must act, but without

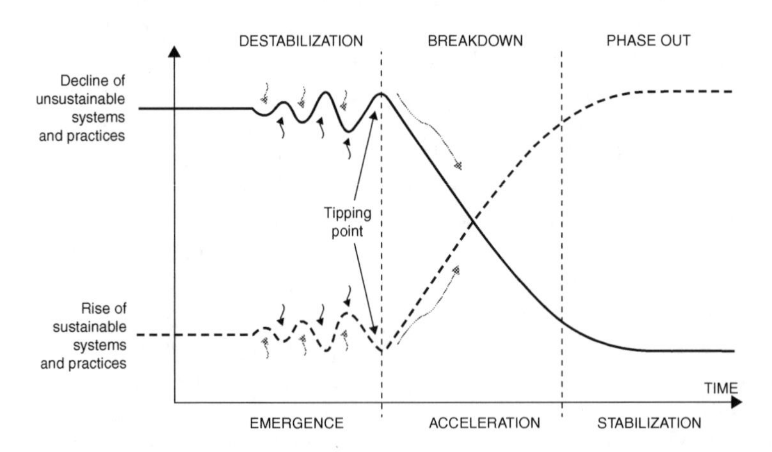

Figure 6.1 Three phases of transformation

Source: adapted from ISG (2023, p. 65)

ignoring the need to understand the complexity of society. It is our hope that this book can reveal the complexity of social change in a way that facilitates rather than hinders action.

We begin by presenting what is meant by social transformation. In the current discussion, transformation is frequently mentioned, often in a rather imprecise or undemanding way, and sometimes in an elitist manner (see Box 5.5). There is a great need to provide a more informed view of transformation and how it can be initiated and achieved. We present five key principles that serve to safeguard its radical meaning. We then explain in concrete terms what social transformation can mean in relation to the five facets of the social. We also emphasize that social transformation is inherently related to power, and that a power perspective is always important when working for social transformation. In the final section, we return to the basic rationale of this book: the current crisis and its challenges, and the urgent need for societal transformation.

What does transformation mean?

Transformation has become a buzzword in much of the literature on sustainability, as well as in the work of expert organizations to guide environmental action (Lidskog and Sundqvist 2022). The IPCC states that limiting global warming to well below 2°C will require transformative and systemic change, and that fundamental sectors – such as energy, land use, cities, infrastructure (transportation and buildings), and industrial systems – will need to transform rapidly (IPCC 2023). Similarly, IPBES stresses that transformative change is necessary to achieve the 2050 vision for biodiversity (IPBES 2019). Furthermore, the expert assessment of the UN Sustainable Development Goals (SDGs) concludes that the world is far from being on track to achieve the SDGs, that incremental and fragmented change is not enough, and that the only way forward is to transform how we think, live, produce, and consume (ISG 2023).

These are examples of how expert organizations assessing and synthesizing the knowledge within an environmental field have gradually come to the conclusion that past decisions and policies have not resulted in sufficient action, and that there is an urgent need for more radical and far-reaching change.

At the same time, there is a risk that the call for transformation will be mere talk, agreed upon but not translated into transformative action. Changing the world is easier said than done, and many proposals for transformation may be unrealistic because they only consider options that may be technically possible, and do not take into account social structures and processes. The world is stratified, with many ways of life, and different groups face different environmental, social, economic, and political challenges (see Chapter 3), as well as different ways of understanding these challenges (see Chapter 4). It is therefore important to have a realistic sociological view of society, not least when aiming for large-scale, systemic change.

Björn Linnér and Victoria Wibeck (2019) have studied social transformation, both theoretically and historically. Societal transformation is multifaceted, non-linear, and complex (see definition in Chapter 1), which they illustrate by investigating historical examples of transformation. They discuss the Industrial Revolution, the abolition of slavery, and automobilization. Numerous factors interact to shape the process of change and the outcome. There are also several counter-factors that can delay or limit transformation. If there is one thing we can learn from historical comparisons, they argue, "it is that transformation is never finished, but is ongoing" (p. 197).

Linnér and Wibeck distinguish between macro-transformations and particular transformations. The former refer to the change of entire civilizations or the rise of new epochs. The Industrial Revolution is a good example, but they also discuss more abstract phenomena such as "modernity" and "the Anthropocene." Particular transformations include changes in infrastructure systems. This could be a shift toward automobilization and away from the use of horses as the main mode of transportation, for example. The invention of the internet is a relevant recent example of such a socio-technological transformation, and today there are intense debates about where artificial intelligence will take us.

Taking into account Linnér and Wibeck's theory, as well as common insights from sociology, we suggest that transformation can be described and understood in a number of ways, as shown in Box 6.1.

Box 6.1 Aspects of social transformation

Scale: Macro-transformation versus particular transformation.

Timeframe and pace: Protracted versus rapid change (Linnér and Wibeck 2019). A similar distinction is made between incremental (step-by-step) and radical change. The idea of "modernization" leans toward the incremental side.

Depth: We can speak of a transition (a passage, just a shift from A to B) or a transformation (a more fundamental change in form; shifting from A to B implies changes in C, D, E…) (see Linnér and Wibeck 2019).

Linearity: Change can be conceived as linear, non-linear, dialectical, exponential, or chaotic. Statistics, graphs, and other forms of visualization can greatly shape (and simplify) how we view change.

Disruption: Change can be seen as requiring the collapse of something that already exists (as emphasized by Karl Marx and Joseph Schumpeter) or the continuing development of something that already exists (as emphasized by Emile Durkheim).

Agents of change: Agents of change can be conceived in terms of "bottom-up" processes (social protest movements, social tipping points, and the mainstreaming of niche experiments), or "top-down" processes (policy change, regulatory reforms, infrastructural planning, international agreements, military coups). For example, the theory of the risk society and reflexive modernization (see Chapter 2) stresses the importance of a social rationality that is evolving in bottom-up processes, while top-down measures can facilitate such development.

Intentional or unintentional: Change can be planned/intended, such as the abolition of slavery and the invention of democracy, or it can be unplanned/unintended. An example of the latter is disasters that cause irreversible effects and have a huge impact on society. Another example is the gradual change, through societal differentiation, that led to the division of labor that Adam Smith and Emile Durkheim wrote about. Also, an intended change usually has both intended and unintended effects.
Change metaphors: Common metaphors in narratives of transformation include a "journey," a "building process," a "war," "co-creation," and "recuperation" (Linnér and Wibeck 2019). All of these have their specific pros and cons, and they can be combined in various ways. For example, the war metaphor may be apt for describing the fight against a pandemic, but it can also be used to legitimize undemocratic measures. The co-creation metaphor may be useful when talking about the need for collaboration between different groups in society, but can be overly consensus-oriented and underestimate the role of power, resistance, and vested interests.

Categories such as those in Box 6.1 are important to keep in mind when describing and explaining transformative change. Change in one sector or society may not look the same way as in another sector or society.

Social transformation: Structure and agency

In Chapter 2 we stressed the duality of society – that society and human beings are interdependent, inconceivable in isolation, and constantly influence each other. This duality means that human beings both shape and are shaped by society; they develop their thoughts, feelings, desires, and habits in particular social contexts, while at the same time shaping and sometimes changing those contexts. The concept of social structure refers to how our actions and choices are conditioned, while agency refers to the ability of actors (individuals, groups, or organizations) to make decisions, take actions, and change their circumstances. A sociological understanding of society means that social processes at the micro, meso, and macro levels are interdependent and dynamically related. They always affect each other, often in unpredictable ways. We have also emphasized that society is multifaceted and diverse (Chapter 3), that we understand the world in very different ways, and that we therefore also act differently in the world (Chapter 4). This also explains why many people and organizations are critical of, or uninterested in, the idea of changing their practices, and place their hopes in simple, but flawed, solutions (Chapter 5).

This understanding of society means that there is no single cause of all kinds of ecological destruction, such as economics (capitalism), technology (industrialism), politics (nation-states), or culture (worldviews). Instead, all of these and other causes are involved in creating the current environmental challenge. As we have emphasized throughout this book and will discuss further below, all five facets of the social are important in exploring how to transform society. Therefore we must focus on all five facets in order to understand the causes of environmental destruction as well as the options for transforming society toward sustainability.

This has critical implications for how transformative change is initiated and created. In the current discussion, transformative change is often framed and understood as a problem of design, where the solution is an intended, programmatic process of change toward specific objectives. Policies are developed, different sectors are given targets to reach, measures are invented to monitor the work, and periodic evaluations are conducted to ensure that sectors are reaching their targets. If society is designed in a suitable way, if global agreements can be reached and countries and companies abide by them, if technology is mobilized to invent smarter processes and products that are then disseminated globally, and if citizens support this development – then society is expected to get back on track and begin to transform. This is a common way for countries and the international policy community to work. Sometimes it has been successful – the international community and nation-states have achieved some of their goals through this kind of top-down rational planning and work. Sometimes, however, it has been unsuccessful – policies have been formulated but not implemented, or, in the worst case, policies have been directed at goals in the distant future, thereby enabling the continuation of an unsustainable society ("we don't need to reduce our emissions now, we'll do it in a few years"). So this work can be, and often is, important, but it is misleading to think that it is the only way to bring about transformation. If this kind of work is thought to be the only thing that matters, then many options for transformation are made invisible and not considered.

Much of the current discussion on transformation is based on a simplistic, overly rationalistic, and hierarchical understanding of society that emphasizes macrosocial solutions such as international regulations and new technologies that can (and should) be applied globally. There is often an implicit separation between global and local, between political action and everyday activities, between large and small, and between systems and actors. A particular problem with this discussion is that it tends to create a limited view of agency. Indeed, much environmental research tends to underestimate the greater impact of small-scale and local actions (O'Brien et al. 2023). One reason for this is an oversimplified view of how social change occurs, which conceptualizes society as remote from, or unaffected by, many actions. Personal lifestyle changes (such as choosing not to travel by air or eat meat), local actions (such as boycotts and local restrictions on urban road traffic), and even some national actions (such as national gasoline taxes) are disregarded and seen as merely symbolic, while the crucial decisions and actions are taken at the international level by powerful actors, not least large countries and transnational corporations. Social transformation must be understood more broadly, as involving a multiplicity of actors, activities, and sectors, and in terms of power being dispersed. Agency is increasingly constructed in networks and assemblages consisting of people, organizations, norms, and technologies (Lidskog and Standring 2023). This means that it is important to stress the duality of society and to recognize both structural conditions and opportunities for action.

This is exactly what Ulrich Beck (2016) emphasizes in his last book, *The Metamorphosis of the World*. His message in this book is that the current way of understanding the world, with its emphasis on (limited and non-transformative) solutions, blinds us to what is actually taking place today. In his book *Risk Society* (which we discussed in Chapter 2), he stressed the negative side of a positive developmental trend (severe environmental risks caused by the production of material wealth). In *The Metamorphosis of the World*, Beck instead discusses the positive side of a negative trend (opportunities for action caused by the production of severe risks and disasters). According to Beck, the current global crisis – exemplified not least by climate change – makes possible a new understanding of the world. Of course, climate disasters cause great harm and human suffering. But they also allow for a new perspective in which people around the world recognize their mutual interdependence, which in turn creates opportunities for concerted action. In this sense, risk, crisis, and failure can be triggers of agency and change. For example, a climate disaster can devastate a local community, but it can also lead to a broad, transboundary commitment to global justice and strong support for the affected community, coupled with holding the fossil fuel industry and nation-states accountable for the disaster. Risks and disasters can create what Beck calls "transnational risk communities," where actors with different interests can now act together because they are all threatened by the same risks.

The climate crisis offers an opportunity to find new ways of seeing the world, a possibility to look beyond national frameworks and develop a cosmopolitan outlook. Previously important distinctions – such as us/them, local/global, national/international, the micro level of everyday life/the macro level of the big players – are now becoming blurred and obsolete. Beck argues strongly that the prevailing national perspective in politics must be replaced by a cosmopolitan approach, in which organizations and people begin to see themselves as part of, and agents in, a global, cosmopolitan world. Beck even claims that we now have a new way of seeing and being in the world and of doing politics. We need to understand that what was unthinkable yesterday is not only possible today, but often is also real. Society is already changing, and challenging existing ways of thinking and doing can enable people and organizations to act in creative and transformative ways.

While Beck's message may be overstated – there are reasons to doubt that society is already transforming in the ways he describes – it is crucial to recognize that there may be opportunities for transformative action that we cannot see, because of our limited or mistaken understanding of society and of how social change occurs. Many researchers share this view, that new ways of understanding and thinking can lead to new actions, just as new ways of acting can influence our understanding and thinking (see, e.g., Meadows 2008, O'Brien et al. 2023). For example, recognizing that microsocial practices and the larger social order are interconnected and influence each other can counteract the perceived gap between everyday life and global institutions (see, e.g., Boltanski and Thévenot 2006). Thus, by taking a broad view of social transformation, where the power to change society lies in all five facets of the social, one also justifies action. There is a need to promote a broader understanding of social change, a view that action is meaningful no matter who does it, and no matter what level it takes place at, be it micro, meso, or macro. Many actions and interventions can be part of larger processes of social transformation – and inaction and non-intervention can also slow this process.

Five guiding principles

Based on our discussion so far, we would like to suggest five principles for qualifying the meaning of social transformation, and preventing the concept from being emptied of its radical meaning.

First, it is important to adopt a theoretically and methodologically open-ended and pluralistic understanding of transformative change. Historical analyses of previous transformations, such as those studied by Linnér and Wibeck, and categories such as those in Box 6.1, can remind the analyst of this need. Transformations vary widely, and they are multifaceted. The analyst ought to compile insights and perspectives from a variety of perspectives. For example, we think it is more constructive to combine and contrast the macro-theories that we presented in Chapter 2 than to choose one as the *a priori* favorite theory for all kinds of analyses of macro-change. And macro theories can be combined with meso and micro theories, and vice versa.

Secondly, the perspective must be able to move beyond an overly rational (see previous section) or problem-solving mindset (Chapter 5). Problem solving is not inherently wrong; on the contrary, it is essential for human survival and well-being. But to deal with the current civilizational crisis, it is inadequate and potentially misleading to believe in simple and piecemeal solutions that do not involve any fundamental changes in societal activities. Power structures shape our ways of seeing, thinking, and acting in society, which is why it is so important to avoid wishful thinking and to critically evaluate proposals and assess the extent to which they actually will stop unsustainable activities and support sustainable ones. In this way, solving problems and finding solutions can be beneficial, but only if they are part of a broader strategy for transformative change.

Thirdly, the perspective must be able to address the issues of complexity and resistance to change, which includes being able to confront power structures at different levels and vested interests that work to maintain unsustainable practices. Given the planetary scale of the current socio-ecological crisis, the necessary changes will not be free of tensions, conflicts, and compromises. New technologies, practices, and institutions will have to contend with existing ones. Transformative change is to some extent conflictual (Boström et al. 2018), because change necessarily involves winners and losers, and compromises between competing interests. Those who profit from the fossil fuel industry will obviously have something to lose from the transformation to a fossil-free society. Thus, power and its exercise are not necessarily negative things. Indeed, they are necessary for mobilizing and realizing change. Power is an important resource for spreading awareness of the crisis and its causes, for changing common mindsets, and for overcoming obstacles to social transformation.

Fourthly, we need to cultivate a reflexive understanding of the change process, that is, an ongoing, self-critical review of change processes to enable necessary adjustments to be made. This is needed because change is continuous, produces unintended effects, is characterized by trial and error, and will create numerous dilemmas and ambiguities. Such reflexivity must involve both the scientific study of change and the change-makers themselves. (On reflexivity, see Boström et al. 2017, 2024.) Cultivating a reflexive understanding is also important for creating robust transformative processes. History is full of cases where people and organizations have struggled for good causes, but ended up causing failures and disasters (Scott 1998). There is a great need for all actors involved in working for transformative change to be self-critical and to reflect on their own epistemic assumptions, normative commitments, and social embeddedness (Lidskog et al. 2022). When actors place too much faith in their own concerns and capacities, there is a great risk that the concerns and evaluations of other actors will be ignored, denied, or rendered invisible.

Fifthly, transformative change must involve all five facets of the social. This is the core message of this book; society is not a separate entity external to our social life and the way we organize it. Instead, the facets are intertwined, interconnected, and co-constituted. This means that social agents – actors with some form of capacity to change society – can be located "everywhere" in society. Of course, they will have different capacities and opportunities, but our point is that transformative change can begin and develop at the micro, meso and macro levels, and involve all five facets. Therefore, the next section will discuss these five facets to show their importance in initiating and facilitating transformative change. We will return to the topic of solutions, but this time we will focus on solutions that we believe have transformative potential. We will refer both to existing examples of people, movements, practices, and infrastructural changes that are attempting to change society and to various theories related to transformative change.

Box 6.2 Five principles for studying transformation

1. Adopt a theoretically and methodologically open-ended and pluralistic understanding of transformative change.
2. Go beyond the problem-solving mindset.
3. Address complexity and resistance to change.
4. Cultivate a reflexive understanding of the change process.
5. Involve all five facets of the social.

Social transformation and the five facets of the social

Power permeates all five facets of the social. As shown in Chapter 1, power is multidimensional, and it enables certain actions and hinders others. It is crucial to see that power is linked to agency (see Chapter 1). This includes the mobilization of counter-power, a form of power that can challenge and confront existing unsustainable structures, dominant groups, and ways of life. It is therefore a question of empowerment, which can take place at personal, interpersonal, and more collective levels.

Firstly, it is a matter of mobilizing resources or, we might say, power resources. Like Michel Foucault and Anthony Giddens, we might think of power in terms of *transformative capacity*, "the ability to intervene in a given set of events so as in some way to alter them" (Giddens 1987, p. 7). For this to happen, the agent needs to have resources. While Giddens stressed allocative resources (money, material facilities, technology) and authoritative resources (e.g., legitimate power within a specific domain), Foucault emphasized the important role of knowledge and discursive power (1980). In relation to facet two below, we will speak of the role of social power resources or, in other words, social capital.

Secondly, power includes the ability to set the agenda, to control what issues are discussed and decided upon (Lukes 2005). Much power is exercised by either putting an issue on the agenda or keeping it off. Actors working for transformative change must put the need for transformation on the agenda and counter attempts by other actors to prevent it from becoming a part of political and public discourse.

Thirdly, it is about resisting manipulation and taking control of the message (Chapter 4). It is meaningless to put social transformation on the agenda if it is stripped of its radical meaning and addressed only within a problem-solving mindset (Chapter 5), or if impossibilism and a sense of despair develop, leading actors not to take action and demand change even when they believe it is needed.

These aspects of power are important for making people aware of a problem, for enabling counter-movements to arise, and for demanding and initiating processes of social change. In the following, we will investigate how all five of the facets are involved in bringing about transformative change. The discussion will not be exhaustive, but rather aims to show how the facets matter, and that power is an important aspect of this work. Alongside power, it is also important to consider time and space. Struggles for transformation will concern different issues and have different characters because of their historical and geographical contexts.

Change in the inner life

Change must and will involve people. Although this is obvious, it is often forgotten. People both adapt to change and are initiators of change. People will change no matter what, either because they have to adapt to changed circumstances or because they can and want to – or a combination of both.

People may participate in climate protests. They may get involved in civil society organizations, especially environmental movements. They may engage in efforts to reduce their own climate and ecological footprints through lifestyle changes. They may try to influence others in conversations, or they may be encouraged by the exemplary words and actions of others. They may serve in organizations as colleagues and experts, and suggest ways to create more sustainable activities in the organizations.

Regardless of how change will happen, a major transformation or even disruption of society such as we foresee cannot be achieved without a high degree of

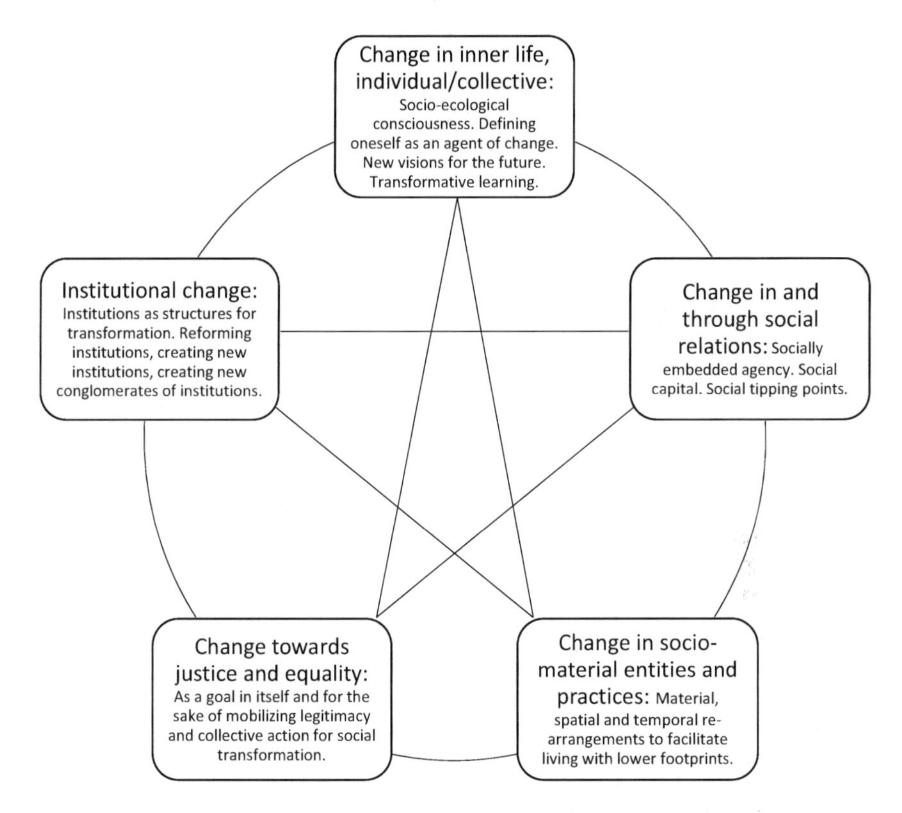

Figure 6.2 The five facets of the social and social transformation

awareness-raising. This cannot be achieved through simple and ordinary information campaigns (see our discussion of cognitive fixes in Chapter 4). Instead, change must come through a new *socio-ecological consciousness*, a much greater, deeper, and more holistic awareness of how our lives are conditioned by society-nature relationships, and how these conditions have become destructive. Thus, it is crucial to have a relational understanding of transformation, that everything we do matters, and that we are all part of the processes that work for or against transformation.

This will not be an easy task. In Chapter 2, we discussed why we do not live in an ecologically sustainable way and gave several reasons at the micro-social level for why we are failing to take the long-term future into account. Put simply, social aspects take precedence. We focus on the "here and now" in the immediate context: relationships, social status, convenience, needs, and the social demands of the day. Immediate goals and ambitions tend to occupy a more prominent place in people's consciousness than long-term protection of the environment. In many contexts, the social is associated with the present and the environmental with the future, and the future is then conveniently forgotten. At the same time, we humans have a unique potential to look far into the future and change our actions based on what can foresee. We have the capacity to do "uncomfortable" things in the present because we expect them to lead to a positive outcome in the future. People exercise, even if they do not like it, to get in better shape. People save money to be able to buy a home. Students study hard to get the job they want. Thus, when we are committed to certain goals and values, we are willing to act to realize them, even if it means enduring hardship along the way.

Awareness and aspirations can become collective. A feature of social movements is that they have developed a collective awareness of the situation and succeeded in getting people to aim for change (Eyerman and Jamison 1991). Since its inception, sociology has spoken of collective subjects becoming conscious of themselves as agents of change. This is about defining oneself and one's peers as a force for historical change. Marx's famous notion of "class consciousness" describes a situation in which people who belong to a class become aware of their situation, mobilize, and work to change social structures (to revolutionize society). From a less political and less normative viewpoint, other sociologists have pointed to the importance of social mobilization and social movements in processes of social change (della Porta and Diano 2020, Tilly 1978). Social movements have played a key role in democratization processes and the construction of welfare societies. This is essentially what Greta Thunberg is talking about when she claims to represent the youth of society and argues that future generations should have something left to live for (Thunberg 2019). Movements like *Fridays for Future* and *Extinction Rebellion* are basically about this.

A protest can be many things, but a key feature of protests is the creative invention of alternative storylines and frames (see Chapter 4) – such as the coining of new terms like "climate justice," "minimalism," "car-free cities," and "flight shame" – that cast existing phenomena in a new light or invent new phenomena altogether. This inventiveness is about anchoring the personal level (inner life) with common agendas, about generating new ideas of what constitutes a good and meaningful life, and engaging in meaningful struggle to improve our common circumstances. These frames enable actors to see problems in a new light, challenge existing patterns, and find ways forward. Social protest movements must engage in such reframing of visions for the future (Benford and Snow 2000).

We have already talked about how resistance to change is related to the inner life (see Chapter 5). But what is it that opens us up to change? While there is no simple answer, an important piece of the puzzle has to do with our emotional capacities. Studies in environmental psychology have found that reflecting on emotions such as discomfort, guilt, anxiety, and frustration does not have to lead to passivity, but can instead lead to "critical emotional awareness" and become a constructive force (Ojala 2016, 2023). Action, as opposed to passivity, can be an important tool for generating hope. In relation to the environment and non-human nature, there are a variety of emotional responses that are key to our actions: our human capacity to feel awe, to be touched, to be shocked, and to be surprised.

Changes in inner life require re-consideration, re-evaluation, re-visioning, re-thinking, and re-learning. Many things that have been taken for granted in our everyday lives now need to be questioned. These include basic assumptions and worldviews, such as assumptions about what leads to progress, development, and quality of life. When we fundamentally change our perspectives and frames of references in this way, we can speak of transformative learning (Boström et al. 2018, Boström 2023, Mezirow 2009).

In cases where people want to reduce their ecological and climate footprint (Boström 2022), learn to become vegans (McDonald 2000), or become environmental activists (Burt 2019, Kovan and Dirkx 2003, Moyer et al. 2016), the transformative learning perspective has been shown to be suitable for describing the profound lifestyle changes that are required.

For example, people who want to get into the habit of using a bicycle instead of a car in everyday life may not need to learn *how* to ride a bike. That part is easy. Rather, the transformative aspects have to do with reevaluating the meanings of driving and biking, which involves dealing with questions such as when, how far, and how often they are able and willing to take the bike instead of the car. It may involve learning more about the practical and safety aspects of bikes and cycling, but can also be about more abstract issues such as identifying and exposing the dominance and normalcy of mass motorization, critically scrutinizing the car norm, discovering how embedded it is in social life and in gender relations, observing how strongly it shapes policy, and realizing how cycling infrastructure is often forced to adapt to this norm. It is about finding ways of dealing with the car norm, finding ways to express criticism, handling value conflicts, and defending cycling as an alternative. It is also about evaluating and appreciating how quality of life can be improved through active mobility practices such as cycling, perhaps with reference to positive role models. This transformative learning has to do with changes in the inner life (our assumptions, understandings, values, emotions, norms, and sense of identity).[1] You can replace the example of cycling with reducing meat consumption, becoming a vegan or vegetarian, ceasing to buy new clothes, adopting minimalism, or downsizing one's work and consumption. Think about these examples and reflect on the changes in your frames of reference that such lifestyle changes would likely entail. Also, do you think your life would be better or worse if you changed it in this way?

This may seem to be focused on the individual level, but it is crucial to reiterate that transformative learning is both individual and collective. We change the inner life together; it is social. Groups and communities of people can also be seen as having a collective inner life, collective identities, and engaging together in processes of transformative learning (on the collective dimension of transformative learning, see e.g., Taylor and Cranton 2012, Pisters et al. 2019, Buechner et al. 2020).

In summary, crucial questions to ask are: How can we create and support situations and contexts that facilitate personal and organizational reflection and inner-life change? How can we support the ability to pay more attention to non-material values and the abandonment of many material values that lead neither to personal well-being nor to a sustainable future? What are the common frames of reference (visions) and role models (positive reference persons or collectives that provide new identities) that can stimulate shared reflections, actions, protests, and transformative learning to work for a sustainable future?

Change in and through social relations

Everyday life is socially embedded. Therefore, when thinking about the agency behind change, we should avoid using an overly individualized notion of change, and instead think in terms of social agency, a socially embedded agency. In this section, we will make three related points. First, change occurs in and through social relations. Secondly, social relations and the wider community network serve as resources in change-related work. Thirdly, major processes of change can spread through the network of social relations and interactions, leading to the possibility of social tipping points.

So first, whatever form of change we are discussing in everyday life, it is essential to take social relations into account. This is no less true when we consider transformative change. Again, we can take the example of a major lifestyle change, such as downsizing or going vegan. Research shows that social relationships are important for such lifestyle changes to occur. We should not expect such change to happen individually; it usually requires a supportive social environment. If a person wishes to begin a process of transformative lifestyle change or wants to become a member of an environmental protest movement, the support from immediate interpersonal relationships can be critical. Such support may involve practical, cognitive, and emotional aspects. Receiving moral support from like-minded people can be particularly important if the person feels that the change process deviates from prevailing norms in society, such as those associated with patterns of overconsumption (Boström 2023). Not only are one's closest relations important, but collaboration, conversation, and support involving the broader social networks of friends, relatives, and neighbors can also be crucial. Boström's empirical studies of reduced consumption practitioners show that people support, discuss, debate, negotiate, and try out activities together with their social circles (families, friends, relatives, etc.) (Boström 2021). Such joint efforts can also lead to positive changes in the relationships, deepening them.

Secondly, a prerequisite for channeling change through civil society groups is the mobilization of social resources, understood here as the connections between members of these groups. In social science, this is known as "social capital," the social network that surrounds an actor or the character of relationships within a community. When an actor – a person or an organization – has access to social capital (a rich social network), this resource can enhance the action capacity to work with various issues, including the capacity to initiate transformative change in society.

For example, in both rural and urban areas around the world, we can see many new bottom-up community transformation initiatives, where groups of committed people come together to collectively develop more sustainable lifestyles and higher levels of self-sufficiency (see, e.g., Schor and Thompson 2014, Kallis et al. 2018, Carnevale et al. 2023). These initiatives may go by different names: degrowth movements, convivialist movements, sharing communities, cycling communities, and so on. They may focus on sharing things and resources, achieving shared consumption, cultivating skills (e.g., cycling, repairing, sewing, crafting, learning what is edible in the forest). They may focus on joint production, such as growing vegetables in community gardens or sharing ownership of a renewable energy facility, or they may provide various services to the local community. In such civil society activities, social capital is always a key resource. Along with social capital comes the strengthening of trust relationships. Trust makes it easier to share valuable goods with others and to initiate new activities. Social capital can also be critical in helping people cope with a variety of situations, including those characterized by danger and uncertainty, such as disasters (see Chapter 3).

We can speak of "communities of transformation" when discussing cases like the examples mentioned above. An interesting example is the *transition town movement*. Studies of this movement and similar initiatives often point to their inclusive, open, participatory, and socially empowering spirit (Connors and McDonald 2010, Middlemiss and Parrish 2010, Bay 2013, Feola and Nunes 2014, Barr and Pollard 2017). These movements focus on lifestyle change in the community. They seek to support each other through social activities and by exchanging resources, services, and knowledge. Emotional attachment to a particular place is crucial. In terms of social relations and interaction, a very important feature of these communities is that they try to foster a collaborative atmosphere among the members.

Some transition town initiatives have fostered collaboration with a range of actors in the community, including local businesses and media, social enterprises, non-governmental organizations, and other transition initiatives in the region or further afield (see Feola and Nunes 2014). In this way, social capital can be strengthened not only between people in a community, but also between organizations in a larger context.

Such examples are not without challenges. They may struggle with internal conflicts and ideological divisions or be difficult to scale up. Can examples like these inspire and spill over to many more people, groups, and communities around the world and become more mainstream in society? We will relate this question to an interesting idea about social tipping points as the third point we address in this section.

The idea of social tipping points has been introduced to suggest the possibility that societies can change in a positive direction. This is similar to the notion of "Earth system tipping points," which natural scientists fear could lead to "Hothouse Earth" conditions, where parts of the world become uninhabitable because of climate change (Steffen et al. 2018). Is it possible to apply such a concept of nonlinear and dynamic processes of transformation to society as well, but to describe processes that work in the opposite direction, toward sustainable transformation? It may indeed be plausible to think in terms of social tipping points, but we recommend being cautious about importing such concepts from the natural sciences (Milkoreit 2023). Social tipping points refer to a particular type of change process where, after reaching a critical threshold, the process develops rapidly toward systemic change (transformation).

Some argue that a number of historical developments can be partly understood in terms of the dynamics of social tipping points, for instance the spread of Protestantism, the abolition of slavery, and the growth of labor unions (Otto et al. 2020). We can also point to the radical shifts in norms around gender relations, sexuality, drug use, and recycling that have occurred in many countries. To a considerable extent, these examples of change involve "contagious" processes of diffusion of social norms, ideas, ideologies, opinions, knowledge, and behaviors.

The theorizing around social tipping points conceives of societies as complex social networks. Ideas, norms, and opinions can spread through *social relationships and interactions* within and across these networks. One reason why they can spread like an infectious disease is that people tend to imitate each other. Imitation includes copying everything from gestures and body language to cognitive and emotional habits. Such mimicking is done unconsciously because of the need to "fit in" in various social settings, to facilitate interpersonal interaction, and to develop social bonds and solidarity. Experimental research shows that unsustainable behaviors can spread through unconscious imitation (Zorell 2020). The same logic can apply to the opposite: sustainable habits.

People can also engage in more conscious opinion formation, and deliberately influence each other in a chain-like manner through the medium of conversation. A study by Zorell and Denk (2021) showed that interpersonal influence through conversation is an important predictor of engagement in political consumerism (e.g., adoption of green and ethical consumption). Thus, both unconscious and conscious processes are involved in the spread of ideas, norms, and behaviors through networks.

When speaking of society rather than natural phenomena, we can consider that each individual, group, organization, and community has its "critical threshold" of conditions where it is likely to change (habits, opinions, etc.). Several factors can influence this willingness to change, such as the character of the social network, or how much one trusts the information provider – the person or actor ahead of one in the chain – or the strength of existing beliefs and habits.

The nature of the issue is also important to consider. The critical threshold for comprehensive behavioral change is likely to be higher than for changes in attitudes and opinions, even if opinions can also be very resistant to change because of socialization, self-identity, social belonging, and cultural values in the society. Often, this kind of resistance involves ignoring and questioning established knowledge (Klintman 2019). Resistance may also be caused by existing power structures and vested interests (see Chapters 2 and 5). Established institutions are inherently inert and have stabilizing mechanisms that oppose change (Otto et al. 2020). Therefore, we should not hope that social tipping dynamics alone can bring about transformative change. Nevertheless, the spreading of opinions, ideas, and habits in social relationships and networks through mechanisms of imitation and conversation is a necessary and important part of a larger change process.

It is also important to emphasize that social tipping points cannot be treated exactly like natural tipping points, because human actions are less predictable (Bentley et al. 2014, Otto et al. 2020). Human beings have the capacity to respond reflectively to external information, such as knowledge about climate change and even knowledge about possible natural and social tipping points. The phenomenon of the "self-fulfilling prophesy" and the role of reputation can be brought into this discussion. A reputation might be based on false information, yet end up becoming true through its consequences. The classic example of this is a bank. If people believe that the bank is insolvent, even if this is not true, individuals and investors may withdraw their money, driving the bank into bankruptcy precisely because of this reputation. On the other hand, if more people begin to believe in the possibility that society can change as a result of social tipping point dynamics, they may change their opinions and behaviors, and put pressure on other people and organizations (including governments and corporations). This can lead to actual change as a result of this belief. Belief in social tipping points can therefore cause them to occur, which in turn can lead to social change (just as belief in their impossibility can block change and lead to the perpetuation of unsustainable practices; see Chapter 5).

In summary, the crucial questions to ask are: What kinds of social relationships and interactions will motivate people to engage in lifestyle change and environmental activism? How can communities facilitate the strengthening of social networks that in turn will generate the social capital needed for transformative action? What cultural and structural features will facilitate such relationships, interactions, and networks? How can we begin to shape and expand new and more sustainable forms of social interaction, to facilitate communities of transformation? How can we create positive social tipping dynamics?

Changes in socio-material entities and related social practices

Caring, cooking, eating, cleaning, working, commuting, vacationing, exercising, sleeping, buying, selling, producing, sharing, communicating, celebrating, socializing, discarding, recycling, learning, and conversing are some of the many social practices of everyday domestic life. All of them require material, spatial, and temporal arrangements, at home, at work, and in places where you spend a lot of time. They require infrastructure, synchronization, tools, and practical skills. Ask yourself, what resources do you need to engage in these practices? How does your home community facilitate them in terms of socio-material infrastructure? If you wanted to change some of your practices to be more sustainable and less dependent on large quantities of things, energy, water, and other resources, would this be possible? Or do you feel trapped in your existing socio-material circumstances?

Change and transformation have to happen somewhere. Everyone lives somewhere, and one cannot wish away one's physical and social environment, including the tools and resources that one uses to live and survive.

Material, spatial, and temporal circumstances make it difficult to change social practices (see Chapters 2 and 5). In the short term, people may feel trapped by their community circumstances. In the longer term, communities can work to achieve lifestyles that allow residents to live well with a reasonable climate/ecological footprint, minimize vulnerability, foster resilience, and become less dependent on large quantities of goods, energy, water, and other resources.

Material entities and infrastructures do more than just provide a context of constraints and opportunities for human action. Theories such as social practice theory, actor-network theory, science and technology studies, and the material culture perspective (see Chapters 2 and 4) all agree that socio-material entities and infrastructures play an active role in shaping people. A prominent current example is the digitalization of society. Society and social life are highly dependent on digital infrastructures. Socio-material entities and infrastructures shape what we are able to do. They shape our understandings, what we take for granted in society, our conception of normality. They activate us and provide us with opportunities, but they also blind us to many possibilities.

On the one hand, existing facilities and infrastructures tend to normalize what already exists. Technologies of heating and cooling, for example, normalize certain expectations about indoor temperature, technologies for cleaning and washing influence conventions around cleanliness, and so on (Shove 2003). We tend to form our understandings (see Chapter 4) on the basis of what exists, which can make it difficult to see the ways in which structures can be transformed (see Chapter 5 on resistance). Our imagination is thwarted. For example, we may take it for granted that cities should be full of cars and parking lots simply because they already are, and because we are unable to imagine the possibility of car-free cities. On the other hand, existing socio-material infrastructures may be perceived as somewhat more tangible, and accordingly changeable, than more abstract social phenomena such as institutions (see below). It might be easier to think about how we can rebuild a city to make it car free, for example, or to replace fossil fuels with fossil-free energy, than to imagine how to transform a global institution such as capitalism. Focusing on a particular place where transformation could occur might be one way to avoid being overwhelmed by the complexity of things (see Chapter 5).

It will be necessary to reconfigure socio-material infrastructures to bring about lifestyle changes aimed at reducing ecological and climate footprints (or, in poor regions, achieving the opposite). Consider, for example, how the social practices listed at the beginning of this section could be realized through means other than the ownership of material goods; that is, through close access to community services, or by normalizing the sharing (rather than buying) of goods and services. The sharing economy requires an infrastructure (digital and/or physical) to encourage people to share resources, and someone will have to create and maintain it. If such an infrastructure exists, the practice of sharing will eventually be normalized. The same goes for the practices of repair, reuse, and so on. Cycling communities need space to move around the area conveniently and safely. Concepts such as car-free cities and 15-minute cities (Jiang et al. 2021) can be inspiring visions and provide storylines and frames (Chapter 4) for thinking differently about the material surroundings. Such concepts not only challenge the car norm, but also highlight the benefits of active mobility, decentralized provision of goods and services, shorter supply chains, green and blue areas in cities, and close access to health care and schools. How can people live well in a more efficient or less resource-intensive way?

High hopes are currently placed on digital infrastructures. Digitalization is often seen as environmentally beneficial because it is expected to reduce the need for transportation. This may be the case, but we need to be cautious about having such hopes (see Freire-Gonzalez and Vivanco 2020, Jiang et al. 2021). First, maintaining the digital infrastructure requires huge amounts of energy and other resources. Between 2010 and 2018, the energy demand for computing services increased by 550% and it is estimated to account for 1% of global electricity consumption (IPCC 2022, p. 140). Secondly, people may take advantage of the opportunities afforded by digitalization while still using their cars for other purposes than commuting (such as shopping or vacationing). Thirdly, the internet can boost drivers of unsustainable mass or excess consumption in a variety of ways (Boström 2023). Fourthly, new information and communication technologies, such as social media, can be used both by actors who are fighting for change and those who resist change. The lesson from all of this is that we should avoid assuming that any particular socio-material entity is inherently environmentally beneficial, and should always be cautious when examining how socio-material entities are used and what their social and environmental implications may be.

In the previous section, we discussed many examples in society, in both rural and urban areas, where groups of committed people come together in a community to jointly develop more sustainable lifestyles. For example, Juliet Schor and Craig Thompson (2014) highlight several innovative and heterogeneous experiments in the social landscapes of different countries. These case studies reveal the importance of the socio-material entities in the local area. The authors use the term "plenitude" to characterize these alternative economies. Plenitude consists of four principles that add up to an alternative lifestyle within a local area. The first is transitioning to fewer working hours in paid employment, which helps free up time to work on slower and more sustainable forms of provisioning. A second is high-tech self-provisioning, including an attitude of being able to make and do things oneself. High-tech includes new technological innovations, such as renewable energy, or smart use of the internet to enable new social arrangements of production, distribution and exchange, and consumption. This can include time banks or platforms for sharing and borrowing goods. A third category is "true materialism." Although the word materialism is commonly associated with the pursuit of high income, luxury, and status consumption, the argument here is that downshifting to low levels of consumption promotes a truer kind of materialism. Doing labor-intensive work and owning fewer goods encourages a more time-intensive and quality-conscious relationship with those few possessions than what is commonly seen in the "throwaway society." The final principle is to invest heavily in the local community, to foster strong social capital and trusting relationships within the community (see the discussion of facet two above). Accordingly, this theory highlights the critical role of actively working in the local context to change socio-material entities and related practices.

These communities of transformation can be seen as laboratories of innovation. They are creative and develop new cultural knowledge and human connections that can potentially be transmitted and reinforced on a larger scale, although this also requires attending to the other facets of the social, especially broad institutional change (see below).

In summary, the crucial questions to ask are: How can we learn to engage with material entities in new, more caring ways, and integrate them into the various social practices of everyday life? How can the democratic management of local communities provide recognition, protective spaces, and material support to various civil society initiatives or alternative non-growth movements? What infrastructures need to be invented to make a more sustainable life possible? What forms of community support are needed for people to live without cars? And, leading to the next facet, given the fact of social stratification, will everyone in the community have equal opportunities to use the infrastructure?

Change toward justice and equality

Any serious attempt at social transformation must consider social stratification. It must take into account the (growing) inequalities and injustices in contemporary society, both within and between countries. It must offer a solution to the problems of poverty and lack of inclusion. This is above all a moral imperative. The eradication of poverty and creation of a just global society that also considers the welfare of future generations and non-human animals are fundamental goals in themselves. In addition, undertaking such tasks is crucial to being able to adequately deal with the environmental crisis we face. In Chapter 3, we listed five reasons why inequality must be tackled. We summarize them again here: (1) inequality (through imitation of the wealthy) spurs overly individualized, status-oriented, and excessively consumerist lifestyles; (2) it destroys trusting relationships, collaboration, and a sense of shared responsibility, and thus prevents collective environmental action; (3) it diverts attention from environmental problems and keeps them off the agenda; (4) it impedes the public legitimacy that environmental policy and politics need; and (5) it prevents international collaboration and communication. "Just transformation" is more than just, it is also a more effective and even necessary means for achieving transformation (see also Bennett et al. 2019, Linnér and Wibeck 2019, pp. 159–160, Wilkinson and Picket 2011).

The need to address social stratification applies at every scale: from local communities to international relations. Large and growing perceptions of injustice and lack of inclusion make it more difficult for people, groups, and countries to trust each other, to reach agreements, to motivate action, and to mobilize broad support and participation across class, gender, age, ethnic, and religious lines.

Therefore, transformation must be linked to equity and social justice, and must take into account the issue of intersectionality (see Chapter 3). This is also one of the reasons why a fixation on technology, information, market solutions, and soft policies is insufficient and problematic (see Chapters 4 and 5). Both the environmental sciences and policy have so far been too silent about social stratification and escalating inequalities. This calls for a critique of theories, discourses, and concepts that neglect the social distribution of environmental problems. There are signs, however, that the alarming topics of poverty and escalating inequalities are being taken more seriously, at least in the interdisciplinary academic debate. It is promising, for example, that the global think-tank Earth4All, composed of leading scientists and economists, includes among its five "extraordinary turnarounds" three that relate to social stratification: eradicating poverty, counteracting escalating inequalities, and strengthening women's participation and empowerment (Dixson-Declève et al. 2022). Many concrete solutions to these challenges will require radical institutional change (see the next section).

The analysis of how social stratification relates to social transformation needs to address not only issues of inequality and injustice, but also how people can be empowered, represented, and directly involved in the work of changing societies and communities. Factors such as gender, education, income, and ethnicity often play a key role in engagement. A person struggling to put food on the table may not prioritize environmental activism. This is also a question of democracy, forms of representation, participation, and deliberation, and as discussed in Chapter 3, attention must also be given to representing constituencies that cannot express their concerns. These include non-human animals, children, future generations, non-citizen residents, and people who live outside a democratic electorate. This is also a case of mobilizing counter-power and social movements. Social movements have always been critical in the work toward transformative change. There have been anti-slavery movements, labor movements, feminist movements, civil rights movements, animal rights movements, and so on. If you look at history and think counterfactually, would there have been change in any of the areas of social, human, and animal rights, and would there have been progress in democracy, working conditions, women's rights, animal rights, anti-racism, and sexual freedom, if there had not been broad and massive protest movements offering alternative ideas for a better society?

Indeed, we should not forget that environmental movements have been quite significant in recent decades, though not strong enough to achieve major transformative change. Similarly, the environmental justice perspective (and climate justice) that we discussed in Chapter 3 is not just a theory; it is also an international movement that actively opposes the forces of socio-ecological exploitation. The dominant powers that environmental justice movements confront are often extremely strong because they are equipped with the considerable economic and political power that global capitalism can provide. Leaving aside the larger quest for large-scale and long-term transformative change, winning even small battles in the local community can be extremely difficult. Organized and persistent efforts are usually required to have a chance in the battle. Looking at the dimensions of power introduced in Chapter 2, some economic and political actors have enormous amounts of power, including resources and privileged access to decision-making arenas (the first power dimension), the ability to remove an issue from the political agenda (the second power dimension), and the ability to influence the preferences of the population in the areas to be exploited with the promise of jobs, trade, and economic prosperity (the third power dimension).

Therefore, public protest and resistance, whether local, national, or global, needs to engage all three dimensions of power. Mobilizing transformative capacity involves first mobilizing the necessary resources and gaining access to decision-making arenas. This is the level of manifest conflict. The resistance needs to mobilize a variety of counter-resources. This includes drawing on social capital in the community and making efforts to mobilize more people for the cause, finding allies, financing the struggle, using and developing relevant environmental knowledge and perspectives, and gaining public recognition for the movement in various media. There needs to be a framing of the issue that attracts broader segments of the population. Mobilizing public support is essential because people are workers, consumers, caretakers, voters, and taxpayers. Numbers count in many ways, not least symbolically. The resistance needs to show that many people support the cause. Resistance needs to be organized in order to access, control, coordinate, and sustain the various resources and to continue campaigning.

Resistance must take into account the second dimension of power. This means not taking for granted the existing agenda that is set by someone else, nor (if possible) simply accepting the rules of the game within the existing institutional structure. Attention must be paid to whether issues on the agenda are deprioritized, put on a waiting list, or even removed from the agenda by a decision that further (and often time-consuming) investigations must be performed before a final decision can be made. Other questions include whether the channels of participation and representation are adequate, whether the people affected by a decision are part of the decision-making process, and whether there are any biases in terms of gender, age, ethnicity, class, or other important socio-demographic categories.

Finally, resistance needs to consider the third dimension of power. This involves developing awareness of key assumptions, values, and worldviews about the issue, and about means of persuasion. Essentially, this has to do with issues related to facet one, our inner life, which concerns who and what influences our ways of thinking, feeling, and imagining.

In summary, crucial questions to ask are: What opportunities exist for promoting equality, empowerment, inclusion, and democracy in a community or society? And how can such efforts be combined with working to improve the environment? What are the conditions for mobilizing social capital and other resources in the community for social movements (the first power dimension)? Who sets the agenda in the community, and which actors are able to articulate agendas? Who is able to structure the conditions for action, influence, critique, and interaction in the local setting (the second power dimension)? Think of a local conflict or issue: What socio-environmental frames shape the understanding of the conflict or issue, and to what extent are these frames critically evaluated in terms of their basic assumptions and worldviews (the third power dimension)? Social stratification is closely linked to the operation of overarching institutions and power structures such as capitalism, democracy, and the welfare state. Let us turn our attention to the last facet.

Institutional change

Questions of institutional change are often deeply controversial. Which of our institutions need to change? Just a few, or all of them? How, and how much? Rapidly or gradually? How deeply? What mindsets (ideologies and visions of what constitutes a good society) should guide the change? Do we need new institutions, or can we use existing ones to transform society? Can we use some institutions (science, democracy, media) to reform or revolutionize others (the economic system, political systems)? Or are they so intertwined that only a widespread, popular, bottom-up uprising can change the current order?

In Chapter 2, we introduced some macro theories that are similar in some respects, but also very different in how they interpret change and view the need for institutional change. While the theory of ecological modernization is the one that is most optimistic about the possibility of making existing institutions greener, including the economic system, other theories assume that only fundamental systemic change can bring about sustainable development, for example, replacing capitalism with another economic system.

We believe it is too simplistic to see this as a choice between reform and revolution; institutional change is more complex and multifaceted.[2] Even if, in the long run, society will change fundamentally – as we believe it must – the way forward cannot be characterized as simply replacing certain institutions with others. The process will be complex and marked by struggle and resistance. We suggest that parallel types of institutional change may need to occur simultaneously. From this overarching perspective, we propose three ways of looking at institutional change: (i) how existing institutions can be restructured to facilitate transformative change; (ii) how new institutions can be created to facilitate transformative change; (iii) how the design of institutions (old and new) can be used to develop fundamentally new ways of organizing society.

The first discussion concerns how existing institutions can create a platform and structure for the very transformation itself. This question can be divided into two parts: (i) How can existing institutions be used in working for transformative change? (ii) Do these institutions need to be revised or "upgraded" to become relevant to this work?

These two questions are what many, perhaps most, sustainability debates are about. They focus on what already exists and discuss how to improve these structures. Many discussions of environmental governance and policymaking are of this kind. One question, for example, is how to get organizations and the public to act more sustainably and less harmfully. What roles should different governance mechanisms and regulatory measures (whether legal, financial, informational, educational, and organizational) play in this work? How can public and private actors, including those from civil society, work together in novel ways to create new norms and standards for more sustainable business? How can media institutions better integrate the climate issue into their news reporting? How can environmental education be better integrated into the education system? How can research be changed to better facilitate and guide the work of transforming society?

At the international level, the question is how to use the structures provided by the UN, the European Union, and even large financial institutions such as the International Monetary Fund (IMF), the World Bank, and the World Trade Organization (WTO). The latter are often rightly accused of serving the interests of the economically privileged. Is it possible to reform institutions like these if the economic doctrines change? Dixson-Declève et al. (2022) discuss how poverty could be eradicated by measures such as debt relief for developing countries; limiting or regulating the investments and operations of transnational corporations; creating a new multilateral institution, the International Currency Fund (ICF), to improve the functioning of global financial markets; reforming world trade to reduce the hegemony of neoliberal free trade principles, to support local self-sufficient economies, to protect more sustainable economic development in developing countries, and to take carbon emissions into account in trade relations; and improving access to new green technologies in developing countries. Measures like these cannot be achieved without restructuring the above-mentioned international organizations and establishing new ones.

The current structures for organizing international policies are also criticized for being ill-suited to deal with many international and transboundary problems. The UN, which is composed of countries of which only a handful (the Security Council) possess the power to veto proposals, is often toothless when it comes to conflict-ridden and difficult issues. However, there are ongoing debates and proposals for new ways of organizing international political institutions, ranging from further strengthening the formal way of organizing policy-making (such as a reformed UN) to more cosmopolitan approaches, in which citizens' organizations and social movements are given opportunities to participate in and influence international policymaking (Archibugi 2003, Hassler 2014, Held and Roger 2013). An important component of all these proposals is the creation of institutions that allow citizens to hold not only their own country's political representatives accountable for their actions and inactions, but also those of other countries (as well as other powerful actors who fail to take sufficient action).

This complex of questions also involves the local level. It can include analyses of how different grassroots initiatives, such as transition movements, can be supported by existing institutions in the local contexts in which they operate. For instance, the ability to maintain, spread and scale up communities' experiments with more self-sufficient and sustainable ways of living (see earlier discussion) may require the allocation of resources by an institution such as the welfare state. Gowan and Slocum (2014) found that groups of people that are striving for a self-reliant lifestyle generally require the benefits of a strong welfare state, including access to health care, social security, financial institutions, public transportation, schools, and child care. Similarly, studies of political and ethical consumption and lifestyle change show that the institutional conditions for engaging in such bottom-up movements vary widely around the world. The importance of the quality of the political, educational, democratic, and cultural and media institutions cannot be underestimated (Boström et al. 2019, Karimzadeh and Boström 2022). In addition, a key finding from the study of social movements is that the conditions for mobilizing and pushing for change differ significantly across political contexts. This field of research has used the concept of "political opportunity structure" to compare and explain the conditions under which social movements form, engage in protests, and work for change (Tarrow 2011, della Porta 2022). In very harsh political environments, people may have more reasons to fight for change, but they cannot or will not do so because they would risk being jailed or even killed. Many environmentalists, however, do risk their lives. The murder of environmental activists in Latin America is a well-known phenomenon (Lynch et al. 2018).

This is thus a fundamental issue. The quality of existing institutions is of great importance for social agency and the ability to transform society. Reformulated in our language, we can say that the quality of the institutions (facet five) affects the conditions under which processes of change from the other four facets can be initiated, and how such processes can be maintained, spread, and scaled up.

Another discussion concerns the creation of new institutions that can tackle the pressing issues of our time. Obviously, every institution has a historical origin and trajectory. The question now is: How can new institutions emerge from existing societal contexts? Here we can look at the formation of international collaborations, which, of course, can partly be seen as a task of upgrading existing institutions, as we discussed above, but can also be a fertile area for the creation of new institutions. One of these is the establishment of international collaborations to improve scientific advice. Institutions such as the IPCC and IPBES now exist (but may need to be reformed, see Asayama et al. 2023), while new international collaborations, such as for chemical pollution, are candidates for similar efforts. The increased use of chemicals and novel entities has made them a planetary threat, and pollution is now one of the largest environmental risk factors for disease and premature death worldwide (Fuller et al. 2022, Persson et al. 2022). The severity of the threat has prompted the United Nations Environmental Programme (UNEP) to initiate a process to establish an expert organization similar to the IPCC and IPBES in the areas of chemicals, waste, and pollution prevention. Do we need a world environmental organization at all? Do we need a new United Nations? If so, how likely is the creation of such an organization, given the polarization and stalemate we can see globally? Do we need to create entirely new forms of international collaboration, rather than mimicking existing templates? Or, as Ulrich Beck argues (see our discussion above), is a dramatic reorganization of society already underway, but we do not see it because we habitually view the world as consisting primarily of nation-states through which all change occurs?

Whatever form they take, institutions for international cooperation will have to do more than just investigate and draw conclusions about the natural conditions of the planet. These institutions will have to deal deeply with the world's *macrosocial relations*. It is impossible to disregard international relations, poverty, global injustice, power asymmetries, and the unequal ecological exchange that some of the theories presented in Chapter 2 address. The historical contribution of the rich world to ecological destruction, climate change, and global injustice must be acknowledged, and this is an extraordinarily complex matter with imperialist and colonialist aspects. Institutions – both existing and new – must develop the capacity to recognize and take into account the inherited power structures and privileges in global relations (Armstrong 2019). They must develop the action capacity needed to counteract the powerful vested interests that actively sustain current unsustainable and unjust activities. But how can they do this when these powerful actors also have a firm grip on many of the global institutions and exercise all three dimensions of power?

It may be more likely that interesting new progressive institutions will emerge from below, such as those associated with the transition communities mentioned above. A critical question is: how can such local institutional innovations be sustained and scaled up, when they are easily co-opted and/or destroyed by the business-as-usual economy?

A third discussion related to institutional change has to do with the conglomerate of institutions; in other words, ideas about how the institutions are connected in the transformation processes and in the imagined future sustainable society. This is a promising and visionary new discussion in the environmental and social sciences. A variety of theories and concepts are being debated, all of which discuss fundamentally new ways of organizing society, including degrowth (Kallis et al. 2018, Diesendorf 2020), post-growth (Jackson 2017, 2021), sufficiency (Princen 2005, Jungell-Michelsson and Heikkurinen 2022), the well-being economy (Coscieme et al. 2019), doughnut economics (Raworth 2017), economies of plenitude (Schor and Thompson 2014), economic democracy (Wilkinson and Picket 2011, 2018), the "solidary" mode of living (Brand and Wissen 2021), and Earth4All (Dixson-Declève et al. 2022). They all propose ideas – some more specific, some more abstract – about what a future society could and should look like, and some possible ways to get there. No one believes that change will be easy.

Common to all of these ideas is a critique of economic growth, and in some cases, of capitalism. An increasing number of scholars and debaters are arguing that economic growth cannot continue indefinitely, and there is a growing critique of economic doctrines that reject this understanding. Concepts such as "green growth" are rejected because growth itself is the problem. They may not criticize the fact that some economic activities are growing, such as renewable energy, and they agree with the need to eradicate poverty through local economic development. Instead they criticize growth at the aggregate level. To take one example, Tim Jackson (2017), writing about prosperity without growth, discusses the contours of a post-growth society, and in a recent book he offers a sharp critique of capitalism (2021). A post-growth society would involve a social transformation, initiated by a strong and progressive country, toward a service-based economy focused on care, culture, and craft (Jackson 2017). Thus, Jackson does not foresee a society with entirely new institutions, but rather looks at how some existing institutions – although currently constrained by neoliberal doctrines – can be more effectively used to radically transform other institutions.

The problem, as he sees it, is that the modern economy is structurally dependent on economic growth for its stability. Social stability is based on economic stability, and this is a core problem in a growth-based economy. Jobs and prosperity depend on growth. Stagnation leads to unemployment, growing inequality, and social conflict (see Chapter 3, on distribution). The fact that growth has become necessary to maintain economic and social stability must be confronted head on, Jackson argues. We need a systematic reconstruction of the economy that "offers both meaning and hope to the idea of social progress" (Jackson 2017, p. 140). For example, he emphasizes that businesses could be transformed to provide services (e.g., mobility) rather than material things (e.g., cars), and that work should be linked to "participation" in society or community, with a broad reorientation toward the care, craft, and culture sectors. Moreover, society needs to reframe the concept of investment as a "commitment to the future" rather than a form of short-term economic speculation. Instead of destabilizing the economy, causing financial crises, and undermining prosperity, the progressive state and the financial institutions could cultivate and facilitate investment portfolios to protect natural assets, improve resource efficiency, implement clean, renewable technologies, and build the socio-material infrastructures needed for a more sustainable and enriching social life.

Transforming societies into progressive states with degrowth or post-growth economies and challenging fundamentally capitalist structures is obviously not an easy matter, given the complexity of the task and the powerful resistance to change (Chapter 5). Before global society can overthrow capitalism – if that is what is needed – new economic institutions (based more on circular business models, economic democratization, sharing economies, services like care, craft, and culture, and new local currencies) will have to take shape and grow in parallel. This will take considerable time. The task itself is contradictory; the same institutions that continue to generate the problems will have to be restructured in a process of continuous transformative change, while at the same time room will need to be made for the creation and development of new institutions.

A final reminder is that we must learn from history. With institutional change – and societal transformation more broadly – we can be sure that new, unintended, and unforeseen problems will arise. The process of change will never be complete. Reflexivity – the ability to observe what is happening and not blindly follow a single path forward – will always be needed.

In summary, the crucial questions to ask are: What already existing institutions are reasonably well suited to the task of leading the social transformation? How can these institutions be used to reform or revolutionize other institutions for this purpose? Will new institutions emerge through bottom-up local innovation or in a top-down fashion? How would you describe a sustainable society, and what institutions are needed to get there and sustain it?

A final word

We began this book by emphasizing that society and humanity are under threat. While many countries have highly developed social systems and are lifting people out of poverty, environmental problems are escalating worldwide. Today, the global environmental challenges amount to a civilizational crisis. If nothing is done, we all face the risk of ecological and social collapse. As a result, society must change fundamentally.

Simple solutions, such as technological fixes, will be far from sufficient. We hope we have shown that dealing with the global challenges and the changes they require depends on having a valid understanding of society and how it works. Only then can we develop solutions that are relevant, effective, viable – and transformative. Without an understanding of the social, there is a substantial risk that knowledge will be created, problems defined, regulations designed, and strategies developed that at first glance appear to be beneficial and well thought out, but in fact turn out to be inadequate. We need to act both quickly and reflectively. We will need to re-conceptualize already politically defined problems, such as the problems of climate change and species extinction, to promote a richer view of the social, for example to understand how society is stratified and increasingly unequal, and how agency is socially embedded. And in suggesting ways forward, we need to learn from the history of exploitation and colonialism and not ignorantly perpetuate a history of exploitation of certain people and places.

Calling for transformative change is not wishful thinking. On the contrary, it is wishful thinking to believe that the current trajectories can be sustained. We argue that it is possible to transform society, and to do so in an equitable way. It is a realistic option, but realizing it will require deep commitment and a lot of hard work, persistence, struggle, and conflict. Society is stratified, and the social transformation will not be smooth. However, choosing not to transform society will lead to even worse conflicts in the long run, because continued environmental destruction will result in more and deeper social conflicts.

The task is contradictory in nature. Society is both the platform for change and the target of change, both the problem and the solution. We need to build on existing practices and institutions, while transforming them. We need to critique and transform existing society, while at the same time relying on some existing institutions and resources to provide space and action capacity for the emergence of new institutions and practices that can ultimately replace older ones. We can never escape such apparent paradoxes, but we must learn to live with them with a good dose of reflexivity.

One way or another, social transformation will happen. It cannot be otherwise, because our current trajectories are fundamentally unsustainable. Change may come about through catastrophes or by design, or by a combination of both. What we always do have are opportunities to reflect, plan, and initiate transformative change based on our intentions and thoughts, and building on the current situation. For every day, week, and year that passes while the status quo, resistance to change, and passivity persist, the societal conditions for achieving change will deteriorate.

There are signs that parts of society are becoming increasingly aware of the need for transformation and have begun initiating processes of change. This can be seen in lifestyles, protests, expressions of concern, forms of collaboration, and ways of organizing communities and expertise. We should demand that political and business leaders make more radical decisions, and support those who have already begun to do so. At the same time, we should not believe that only states and large corporations are agents of change that can provide models and visions for an alternative future. Capacities for agency and imagination exist throughout society, even if they are currently often bypassed. The potential for change lies within each and every one of us, in our relationships and in our communities, and we can invent ways of collaborating, organizing, and governing that offer viable alternatives to the current destructive forces that dominate the earth.

Notes

1 For references, see Boström (2022). Regarding quality of life, there are several arguments why materialistic and over-consuming lifestyles do not actually increase quality of life, and may even have the opposite effect (see Boström 2023, Kasser 2017).
2 Some prefer the intermediate concept of "reconfiguration" (see Geels et al. 2015).

References

Archibugi, D. (ed.) (2003) *Debating cosmopolitics.* London: Verso.
Armstrong, C. (2019) *Why global justice matters: Moral progress in a divided world.* Cambridge: Polity.
Asayama, S., De Pryck, K., Beck, S., Cointe, B., Edwards, P.N., Guillemot, H., Gustafsson, K.M., Hartz, F., et al. (2023) 'Three institutional pathways to envision the future of the IPCC', *Nature Climate Change*, 13, pp. 877–880. https://doi.org/10.1038/s41558-023-01780-8
Barr, S. and Pollard, J. (2017) 'Geographies of transition: Narrating environmental activism in an age of climate change and "Peak Oil"', *Environmental Planning A*, 49(1), pp. 47–64. https://doi.org/10.1177/0308518X16663205
Bay, U. (2013) 'Transition town initiatives promoting transformational community change in tackling peak oil and climate change challenges', *Australian Social Work*, 66(2), pp. 171–186. https://doi.org/10.1080/0312407X.2013.781201
Beck, U. (2016) *The metamorphosis of the world.* Cambridge: Polity.
Benford, R.D. and Snow, D.A. (2000) 'Framing processes and social movements: An overview and assessment', *Annual Review of Sociology*, 26(1), pp. 611–639. https://doi.org/10.1146/annurev.soc.26.1.611
Bennett, N.J., Blythe, J., Cisneros-Montemayor, A.M., Singh, G.G., and Sumaila, U.R. (2019) 'Just transformations to sustainability.', *Sustainability*, 11(14), 3881. https://doi.org/10.3390/su11143881
Bentley, R.A., Maddison, E.J., Ranner, P.H., Bissell, J., Caiado, C.C.S., Bhatanacharoen P., Clark, T., Botha, M., et al. (2014) 'Social tipping points and earth systems dynamics', *Frontiers in Environmental Science*, 2, 35. https://doi.org/10.3389/fenvs.2014.00035
Boltanski, L. and Thévenot, L. (2006) *On justification: Economies of worth.* Princeton: Princeton University Press.
Boström, M. (2021) 'Social relations and challenges to consuming less in a mass consumption society', *Sociologisk Forskning*, 58(4), pp. 383–406. https://doi.org/10.37062/sf.58.22818

Boström, M. (2022) 'Lifestyle transformation and reduced consumption: A transformative learning process', *Sozialpolitik*, (1/2022), pp. 1–2. https://doi.org/10.18753/2297-8224-186

Boström, M. (2023) *The social life of unsustainable mass consumption.* Lanham: Lexington Books.

Boström, M., Berg, M. and Lidskog, R. (2024) 'Reflexivity and anti-reflexivity' in C. Overdevest, (ed.) *Edward Elgar encyclopedia of environmental sociology.* Cheltenham, UK: Edward Elgar, pp. 477–482.

Boström, M., Andersson, E., Berg, M., Gustafsson, K., Gustavsson, E., Hysing, E., Lidskog, R., Löfmarck, E., et al. (2018) 'Conditions for transformative learning for sustainable development: A theoretical review and approach', *Sustainability*, 10(12), 4479. https://doi.org/10.3390/su10124479

Boström, M., Lidskog, R. and Uggla, Y. (2017) 'A reflexive look at reflexivity in environmental sociology', *Environmental Sociology*, 3(1), pp. 6–16. https://doi.org/10.1080/23251042.2016.1237336

Boström, M., Micheletti, M. and Oosterveer, P. (2019) (eds.) *The Oxford handbook of political consumerism.* Oxford University Press.

Brand, U. and Wissen, M. (2021) *The imperial mode of living: Everyday life and the ecological crisis of capitalism*. Verso Books.

Buechner, B., Dirkx, J., Dauber Konvisser, Z., Myers, D., and Peleg-Baker, T. (2020) 'From liminality to *Communitas*: The collective dimensions of transformative learning', *Journal of Transformative Education*, 18(2), pp. 1–27. https://doi.org/10.1177/1541344619900881

Burt, J. (2019) 'Research for the people, by the people: The political practice of cognitive justice and transformative learning in environmental social movements', *Sustainability*, 11, 5611. https://doi.org/10.3390/su11205611

Carnevale, A., Pellegrini Masini, G., Klöckner, C.A., Altınay, A.G., Düzel, E., Türeli, B.B., Kerremans, A., Denis, A., and Cacace, M. (2023) *Report on inspiring practice cases.* Trondheim: Norwegian University of Science and Technology. https://doi.org/10.5281/zenodo.7711879 (accessed 5 July 2020).

Connors, P. and McDonald, P. (2010) 'Transitioning communities: Community, participation and the Transition Town movement', *Community Development Journal*, 46(4), pp. 558–572. https://doi.org/10.1093/cdj/bsq014

Coscieme, L., Sutton, P., Mortensen, L.F., Kubiszewski, I., Costanza, R., Trebeck, K., Pulselli, F.M., Giannetti, B.F., and Fioramonti, L. (2019) 'Overcoming the myths of mainstream economics to enable a new wellbeing economy', *Sustainability*, 11, 4374. https://doi.org/10.3390/su11164374

della Porta, D. (2022) 'Political opportunity/political opportunity structure' in D.A. Snow et al. (eds.) *The Wiley-Blackwell encyclopedia of social and political movements.* Chichester: Wiley-Blackwell. https://doi.org/10.1002/9780470674871.wbespm159.pub2

della Porta, D. and Diano, M. (2020) *Social movements: An introduction.* 3rd edn. Malden, MA: Wiley-Blackwell.

Diesendorf, M. (2020) 'COVID-19 and economic recovery in compliance with climate targets', *Global Sustainability*, 3, E36. https://doi.org/10.1017/sus.2020.32

Dixson-Declève, S., Gaffney, O., Ghosh, J., Randers, J., Rockström, J. and Stoknes, P.E. (2022) *Earth for all: A survival guide for humanity*. Gabriola Island, British Columbia: New Society Publishers.

Eyerman, R. and Jamison, A. (1991) *Social movements: A cognitive approach.* Cambridge: Polity Press.

Feola, G. and Nunes R. (2014) 'Success and failure of grassroots innovations for addressing climate change: The case of the transition movement', *Global Environmental Change*, 24, pp. 232–250. https://doi.org/10.1016/j.gloenvcha.2013.11.011

Foucault, M. (1980) *Power/knowledge: Selected interviews and other writings 1972–1977.* Brighton: Harvester Press.

Freire-Gonzalez, J. and Vivanco, D.F. (2020) 'Pandemics and the environmental rebound effect: Reflections from COVID-19', *Environmental and Resources Economics*, 9, pp. 447–51. https://doi.org/10.1007%2Fs10640-020-00448-7

Fuller, R., Landrigan, P.J., Balakrishnan, K., Bathan, G., Bose-O'Reilly, S., Brauer, M., Caravanos, J. et al. (2022) 'Pollution and health: A progress update', *Lancet Planet Health*, 6, pp. 535–47. https://doi.org/10.1016/S2542-5196(22)00090-0

Geels, F.W., McMeekin, A., Mylan, J. and Southerton, D. (2015) 'A critical appraisal of sustainable consumption and production research: The reformist, revolutionary and reconfiguration position', *Global Environmental Change*, 34, pp. 1–12. https://doi.org/10.1016/j.gloenvcha.2015.04.013

Giddens, A. (1987) *Social theory and modern sociology*. Cambridge: Polity in association with Blackwell.

Giddens, A. (1990) *Modernity and self-identity: Self and society in the late modern age.* California: Stanford University Press.

Gowan, T. and Slocum, R. (2014) 'Artisanal production, communal provisioning, and anti-capitalist politics in the Aude, France', in J.B. Schor and C.J. Thompson (eds.) *Sustainable lifestyles and the quest for plenitude: Case studies of the new economy.* New Haven: Yale University Press, pp. 27–62.

Hassler, S. (2014) *Reforming the UN Security Council membership: The illusion of representativeness*. London: Routledge.

Held, D. and Roger, C. (eds.) (2013) *Global governance at risk.* Cambridge: Polity.

IPBES (2019) 'Global assessment report on biodiversity and ecosystem services of the intergovernmental science-policy platform on biodiversity and ecosystem services', eds. E.S. Brondizio, J. Settele, S. Díaz, and H.T. Ngo. Bonn: IPBES secretariat. https://doi.org/10.5281/zenodo.3831673

IPCC (2022) *Climate change 2022: Mitigation of climate change.* Working Group III contribution to the Sixth Assessment Report of the Intergovernmental Panel on Climate Change (AR6). Cambridge, UK: Cambridge University Press. https://doi.org/10.1017/9781009157926

IPCC (2023) *AR6 synthesis report: Climate change 2023.* https://report.ipcc.ch/ar6syr/pdf/IPCC_AR6_SYR_LongerReport.pdf

ISG (2023) *Global Sustainable Development Report 2023: Times of crisis, times of change: Science for accelerating transformations to sustainable development*. Independent Group of Scientists appointed by the Secretary-General. New York: United Nations. https://sdgs.un.org/gsdr/gsdr2023 [downloaded 3 October 2023].

Jackson, T. (2017) *Prosperity without growth: Foundations for the economy of tomorrow.* 2nd edn. London: Routledge.

Jackson, T. (2021) *Post growth: Life after capitalism.* Cambridge: Polity Press.

Jiang, P., Van Fan, Y. and Klemes, J.J. (2021) 'Impacts of COVID-19 on energy demand and consumption: Challenges, lessons and emerging opportunities', *Applied Energy*, 285, pp. 116441–116441. https://doi.org/10.1016/j.apenergy.2021.116441

Jungell-Michelsson, J. and Heikkurinen, P. (2022) 'Sufficiency: A systematic literature review', *Ecological Economics*, 195, p. 107380. https://doi.org/10.1016/j.ecolecon.2022.107380

Kallis, G., Kostakis, V., Lange, S., Muraca, B., Paulson, S., and Schmelzer, M. (2018) 'Research on degrowth', *Annual Review of Environment and Resources*, 43(1), pp. 291–316. https://doi.org/10.1146/annurev-environ-102017-025941

Karimzadeh, S. and Boström, M. (2022) 'Ethical consumption: Why should we understand it as a social practice within a multilevel framework?', *Open Research Europe*, 2, 109. https://doi.org/10.12688/openreseurope.15069.2

Kasser, T. (2017) 'Living both well and sustainably: A review of the literature, with some reflections on future research, interventions and policy', *Philosophical Transactions of the Royal Society A*, 375, 2095. https://doi.org/10.1098/rsta.2016.0369

Klintman, M. (2019) *Knowledge resistance: How we avoid insight from others*. Manchester: Manchester University Press.

Kovan, J.T. and Dirkx, J.M. (2003) '"Being called awake": The role of transformative learning in the lives of environmental activists', *Adult Education Quarterly*, 53(2), pp. 99–118. https://doi.org/10.1177/0741713602238906

Lidskog, R. and Standring, A. (2023) 'Accountability in the environmental crisis: From microsocial practices to moral orders', *Environmental Policy and Governance* 33(6), pp. 583–592. https://doi.org/10.1002/eet.2083

Lidskog, R. and Sundqvist, G. (2022) 'Lost in transformation: The Paris Agreement, the IPCC, and the quest for national transformative change', *Frontiers in Climate*, 4, 906054. https://doi.org/10.3389/fclim.2022.906054

Lidskog, R., Standring, A. and White, J.M. (2022) 'Environmental expertise for social transformation: Roles and responsibilities for social science', *Environmental Sociology*, 8(3), pp. 255–266. https://doi.org/10.1080/23251042.2022.2048237

Linnér, B.O. and Wibeck, V. (2019) *Sustainability transformations: Agents and drivers across society.* Cambridge: Cambridge University Press.

Lukes, S. (2005) *Power: A radical view.* 2nd edn. Basingstoke: Palgrave Macmillan.

Lynch, M.J., Stretesky, P.B. and Long, M.A. (2018) 'Green criminology and native peoples: The treadmill of production and the killing of indigenous environmental activists', *Theoretical Criminology*, 22(3), pp. 318–341. https://doi.org/10.1177/1362480618790982

McDonald, B. (2000) "Once you know something, you can't not know it' An empirical look at becoming vegan', *Society & Animals*, 8(1), pp. 1–23. https://doi.org/10.1163/156853000X00011

Meadows, D.H. (2008) *Thinking in systems: A primer*. Vermont: Chelsea Green Publisher.

Mezirow, J. 2009. 'An overview on transformative learning' in K. Illeris (ed.) *Contemporary theories of learning: Learning theorists... in their own words.* London: Routledge, pp. 90–105.

Middlemiss, L. and Parrish, B.D. (2010) 'Building capacity for low-carbon communities: The role of grassroots initiatives', *Energy Policy*, 38(12), pp. 7559–7566. https://doi.org/10.1016/j.enpol.2009.07.003

Milkoreit, M. (2023) 'Social tipping points everywhere? Patterns and risks of overuse', *WIREs Climate Change*, 14(2), e813. https://doi.org/10.1002/wcc.813

Moyer, J.M., Sinclair, A.J. and Quinn, L. (2016) 'Transitioning to a more sustainable society: Unpacking the role of the learning–action nexus', *International Journal of Lifelong Education*, 35(3), pp. 313–329. https://doi.org/10.1080/02601370.2016.1174746

O'Brien, K., Carmona, R., Gram-Hanssen, I., Hochachka, L.S. and Rosenberg, M. (2023) 'Fractal approaches to scaling transformations to sustainability', *Ambio*, 52, pp. 1448–1461. https://doi.org/10.1007/s13280-023-01873-w

Ojala, M. (2016) 'Facing anxiety in climate change education: From therapeutic practice to hopeful transgressive learning', *Canadian Journal of Environmental Education*, 21, pp. 41–56.

Ojala, M. (2023) 'Climate-change education and critical emotional awareness (CEA): Implications for teacher education', *Educational Philosophy and Theory*, 55(10), pp. 1109–1120. https://doi.org/10.1080/00131857.2022.2081150

Otto, I.M., Donges, J.F., Cremadesc, R., Bhowmik, A., Hewitt, R.J., Lucht, W., Rockström, J., Allerberger, F., et al. (2020) 'Social tipping dynamics for stabilizing earth's climate by 2050', *PNAS*, 117(5), pp. 2354–2365. https://doi.org/10.1073/pnas.1900577117

Persson, L., Almroth, B.M., Collins, C.D., Cornell, S., de With, C.A., Diamond, M.L. et al. (2022) 'Outside the safe operating space of the planetary boundary for novel entities', *Environmental Science & Technology*, 56, pp. 1510–1521. https://doi.org/10.1021/acs.est.1c04158

Pisters, S.R., Vihinen, H. and Figueiredo, E. (2019) 'Placed-based transformative learning: A framework to explore consciousness in sustainability initiatives', *Emotion, Space, and Society*, 32, 100578. https://doi.org/10.1016/j.emospa.2019.04.007

Princen, T. (2005) *The logic of sufficiency*. Cambridge, Mass.: MIT Press.

Raworth, K. (2017) *Doughnut economics: Seven ways to think like a 21st-century economist*. London: Random House Business.

Schor, J.B. and Thompson, C.J. (eds.) (2014) *Sustainable lifestyles and the quest for plenitude: Case studies of the new economy.* New Haven: Yale University Press.

Scott, J.C. (1998) *Seeing like a state: How certain schemes to improve the human condition have failed.* New Haven: Yale University Press.

Shove, E. (2003) *Comfort, cleanliness and convenience: The social organization of normality.* Oxford: Berg.

Steffen, W., Rockström, J., Richardson, K., Lenton, T.M., Folke, C. Liverman, D., Summerhayes, C.P., Barnosky, A.D., et al. (2018) 'Trajectories of the Earth System in the Anthropocene', *PNAS*, 115(33), pp. 8252–8259. https://doi.org/10.1073/pnas.1810141115

Tarrow, S.G. (2011) *Power in movement: Social movements and contentious politics.* 3rd edn. Cambridge: Cambridge University Press.

Taylor, E.W. and Cranton, P. (2012) 'Reflecting back and looking forward' in P. Cranton and E.W. Taylor (eds.) *The handbook of transformative learning: Theory, research and practice*. Hoboken, NY: Wiley & Sons, pp. 555–573.

Thunberg, G. (2019) *No one is too small to make a difference*. London: Penguin Books.

Tilly, C. (1978) *From mobilization to revolution*. New York: Random House.

Wilkinson, R. and Pickett, K. (2011) *The spirit level: Why greater equality makes societies stronger.* London: Bloomsbury.

Wilkinson, R. and Pickett, K. (2018) *The inner level: How more equal societies reduce stress, restore sanity and improve everybody's well-being.* London: Allen Lane.

Zorell, C. (2020) 'Nudges, norms, or just contagion? A theory on influences on the practice of (non-)sustainable behavior', *Sustainability*, 12(24), 10418. https://doi.org/10.3390/su122410418

Zorell, C.V. and Denk, T. (2021) 'Political consumerism and interpersonal discussion patterns', *Scandinavian Political Studies*, 44(4), pp. 392–415. https://doi.org/10.1111/1467-9477.12204

Index

Note: *Italic* page numbers refer to figures and page numbers followed by "n" denote endnotes.

Printed in the United States
by Baker & Taylor Publisher Services